ROUTLEDGE LIBRARY EDITIONS:
SOVIET ECONOMICS

Volume 6

ENERGY REVIEWS: UNIFIED GAS SUPPLY SYSTEM OF THE USSR

ENERGY REVIEWS: UNIFIED GAS SUPPLY SYSTEM OF THE USSR

Edited by
L.A. MELENTIEV

Routledge
Taylor & Francis Group
LONDON AND NEW YORK

First published in 1985 by Harwood Academic Publishers

This edition first published in 2023
by Routledge
4 Park Square, Milton Park, Abingdon, Oxon OX14 4RN

and by Routledge
605 Third Avenue, New York, NY 10158

Routledge is an imprint of the Taylor & Francis Group, an informa business

British Library Cataloguing in Publication Data
A catalogue record for this book is available from the British Library

ISBN: 978-1-032-48466-2 (Set)
ISBN: 978-1-032-49014-4 (Volume 6) (hbk)
ISBN: 978-1-032-49018-2 (Volume 6) (pbk)
ISBN: 978-1-003-39185-2 (Volume 6) (ebk)

DOI: 10.4324/9781003391852

Publisher's Note
The publisher has gone to great lengths to ensure the quality of this reprint but points out that some imperfections in the original copies may be apparent.

Soviet Technology Reviews, Section A

ENERGY REVIEWS
Unified Gas Supply System of the USSR

Volume 2 (1985)

Edited by

L. A. MELENTIEV

Institute of High Temperatures,
USSR Academy of Sciences, Moscow

SOVIET TECHNOLOGY REVIEWS

Soviet Technology Reviews are published under license by OPA Ltd. for Harwood Academic Publishers GmbH.

Harwood Academic Publishers

Poststrasse 22
7000 Chur
Switzerland

P.O. Box 197
London WC2E 9PX
England

58, rue Lhomond
75005 Paris
France

P.O. Box 786
Cooper Station
New York, New York 10276
United States of America

Library of Congress Cataloging in Publication Data

Main entry under title:

Unified gas supply system of the USSR.

(Soviet technology reviews. Section A, Energy reviews, ISSN 0275-7893; v. 2 (1985))
Translated from the Russian.
1. Gas distribution–Soviet Union. 2. Gas, Natural–Soviet Union–Pipe lines. I. Melent'ev, L. A. II. Title: Gas supply system of the USSR. III. Series: Soviet technology reviews. Section A, Energy reviews; v. 2.
TP757.S535 1985 363.6'3'0947 84-29019

ISBN 3-7186-0152-4. ISSN 0275-7893.
Printed in Great Britain by Bell and Bain Ltd., Glasgow.

CONTENTS

PREFACE TO THE SERIES

In the last few years many important developments have taken place in Soviet science and technology which may have not received as much attention as deserved among the international community of scientists because of language problems and circulation perplexities.

In launching this new series of *Soviet Technology Reviews* we are motivated by the desire to make accounts of recent technological advances in the USSR more readily accessible to scientists who do not read Russian. The articles in these volumes are meant to be in the nature of reviews of current developments and are written by Soviet experts in the fields covered. Most of the manuscripts are translated from Russian. In the interest of speedy publication neither the authors nor the volume editors have an opportunity to read proofs, and they are absolved of any responsibility to read proofs, and they are absolved of any responsibility for inaccuracies in the English texts.

Soviet Technology Reviews will appear annually, with the subject areas varying from year to year.

We are much indebted to the volume editors and individual authors for their splendid cooperation in getting these volumes put together and sent to press under considerable time pressure.

The future success of this series depends, of course, on how well it meets the needs and desires of its readers. We therefore earnestly solicit readers' comments and particularly suggestions for topics and authors for future volumes.

By taking this initiative we hope to contribute to the development of scientific and technological cooperation and better understanding among scientists.

THE EDITORS

FOREWORD

The rapidly developing single gas supply system occupies a special place in energy development in the USSR. This system has extensive possibilities for further growth. It encompasses hundreds of gas fields and is connected to consumers by a branching network of interconnected main pipelines, as well as gas distribution networks, scrubbing equipment, underground storage, etc.

This collection of papers includes reviews concerning some important aspects of the development and operation the single gas supply system of the USSR. Particular attention is given to the basic characterstics of this system and some long-term trends in its development.

Soviet experience in solving the problems involved in centralizing gas supply and creating a single centralized gas supply system supplying most consumers will be of interest to the reader.

One of the characteristic features of the gas supply system in the USSR is the fact that many gas-producing regions are far removed from the regions where the gas is used. A powerful construction industry has been created in the USSR to overcome these distances by laying long main pipelines. A special review paper briefly describes this industry.

Soviet experience in improving methods for controlling the single gas supply system could also be of interest. A corresponding single automated control system is in operation in the USSR and a number of subsequent phases in its automation are being developed.

At the same time, general control concepts with step-by-step transitions to adaptive, goal-program, and multigoal (complex) control methods, directed towards the control of efficiency, reliability, adaptivity, and a number of other qualitative system characteristics are also being developed. These problems are also covered in this volume.

In the USSR, the gas supply system is viewed as a subsystem of the energy economy and its development and operation are optimized as part of the total energy balance of the country.

Optimal gas utilization and the basic structure of the single gas supply system are determined from this point of view. The methods used here to construct plans are quite well developed in the USSR and they could be of particular interest to the non-Soviet reader.

The present difficulties in developing the single gas supply system, which are related to the longer distances over which gas is transported, the transition to more complicated geological formations, the more difficult climatic and environmental conditions in regional production, and a number of other factors, are also examined in this volume. The methods adopted in the USSR for overcoming these difficulties are also discussed.

L. A. MELENTIEV

Sov. Tech. Rev. A Energy Reviews, Vol. 2, 1985, pp. 1–52
0275–7893/85/002–001 $30.00/0

THE ROLE OF NATURAL GAS IN THE USSR'S ENERGY DEVELOPMENT

A. A. MAKAROV

Siberian Energy Institute, Ulitsa Lermontova, 130, 664033, Irkutsk, USSR

Abstract

The paper gives an analytic overview of the Soviet publications on the problems of development of the national gas supply system as a part of the overall energy complex. It enters first to the historical role of natural gas in restructuring the energy balance of the Soviet Union during the period from 1960 to 1980; then it describes the methods used for optimizing the development of the gas industry within the energy complex; finally, it gives a comprehensive assessment of the role of natural gas in the future energy strategy for the Soviet Union.

Contents

1. The Role of Natural Gas in the Historical Restructuring of the USSR Energy Balance in the Period 1960–1980

The gas industry is the youngest, most dynamic, and promising sector of the USSR energy complex. Emerging at the end of the 1950s, it has become one of the principal means for the fundamental restructuring of the energy balance of the Soviet Union from essential use of coal and low-grade local fuels (peat, wood, shale, etc.) toward the use of high-grade hydrocarbon fuels (oil and natural gas).

As follows from the data in Table I, with the 280% increase in the total primary energy production during the 60s and 70s and oil production more than quadrupled, natural gas production experienced an almost tenfold increase. As a result, the share of high-grade fuels in the primary energy production increased from 38 to 70%, of which the oil share, from 30 to 44% and the natural gas share, from 8 to 26%. Thus almost nine-tenths of the total growth in the production of primary energy over that period fell on high-grade fuels, of which oil comprised more than half and natural gas more than a third. These figures point to a time technological revolution in the energy complex of the USSR, which resulted in the resolution of two major economic problems.

The first problem was the need to create strong exporting capabilities. Presently gas exports account for about two-fifths of total Soviet exports, (in monetary terms) and, by expanding imports, facilitate more efficient and rapid development of other sectors of the economy. Exports of energy resources increased over the 20 year period by more than a factor of 5.4 and reached the level at approximately one-sixth of the total national primary energy production. Of course, oil resources played the basic role in the solution of this problem.

Further growth of natural gas production to a large extent predetermined the successful solution of the second major economic problem – the fundamental restructuring of internal energy use. To characterize this process, Table II shows the history of natural gas consumption by main users. The table points, first of all, to gradual decrease in the share of electricity generation in gas consumption together with an increase in the fraction of natural gas used to generate heat, where gas is used primarily in small boiler plants. As a result, the total share of gas converted to secondary energy forms exhibits a rather weak decreasing trend, however accounting for about 55% of all gas consumed in the economy.

Direct use of gas (primarily in industrial heaters) comprises the next largest area of gas consumption in the Soviet economy; although its share

Table I Primary Energy Production in the USSR[a]

	1960		1965		1970		1975		1980		1960–1980 growth rates	
	million tons coal equivalent (c.e.)	%	million tons coal equivalent (c.e.)	%	million tons coal equivalent (c.e.)	%	million tons coal equivalent (c.e.)	%	million tons coal equivalent (c.e.)	%	million tons coal equivalent (c.e.)	%
Primary energy production – total	707	100	987	100	1275	100	1600	100	1963	100	1256	100
Oil and condensate	211	30	347	35	504	39	696	43	855	44	644	51
Natural and casing head gas	55	8	153	16	238	19	345	22	513	26	458	36
Total, oil, condensate and gas	266	38	500	51	742	58	1041	65	1368	70	1102	87
Coal	369	52	406	41	440	35	470	29	471	24	102	8
Hydroenergy	27	4	37	4	45	4	42	2.5	61	3	34	3
Nuclear energy					2	~0	7	0.5	22	1	22	2

[a] Compiled from refs. [1], [2] and [5] with updating. Non-commercial energy not included.

Table II Distribution of Network Gas by Use[a]

	1965		1970		1975	
	billion m^3	%	billion m^3	%	billion m^3	%
Electricity production	26.6	21.2	34.8	18.2	39.1	14.9
Heat production	43.2	34.5	69.8	36.6	104.2	39.8
by power plants	16.5	13.2	26.5	13.9	44.7	17.1
by boilers	26.7	21.3	43.3	22.7	59.5	22.7
Total for electricity and heat production	69.8	55.7	104.6	54.8	143.3	54.7
As raw materials	6.8	5.4	14.4	7.6	22.6	8.6
Technological needs	46.4	37.0	68.5	35.9	92.3	35.3
including industrial	36.8	29.4	54.6	28.6	72.6	27.7
Losses	2.3	1.9	3.2	1.7	3.7	1.4
Total	125.3	100.0	190.7	100.0	261.9	100.0

[a]Compiled from ref. [2, p. 198] with updating.

in total gas consumption is somewhat decreasing, it still exceeds 35%. Use of gas as a chemical feedstock is increasing appreciably, but it is still less than 10%.

The changes in the gas consumption patterns together with the considerable growth in the production of natural gas made gas the predominant energy resource for meeting the demands for the so-called boiler–furnace fuel (which includes all primary energy forms except for motor fuels and petrochemical feedstock). This can be seen from the data of Table III: the gas share in the total consumption of boiler–furnace fuel has now reached 37%, with the share of gas as a fuel for power plants and small domestic appliances being more than one-fourth of total fuel consumption, for boiler plants, almost one-half of total fuel consumption, and for industrial furnaces more than 40%. On the whole, gas became the principal boiler–furnace fuel in the Soviet economy.

The shift in the energy balance of the USSR toward increasing the natural-gas share did not occur uniformly throughout the country. As shown in Table IV, the first-turn territories to be supplied with gas were those directly adjacent to the gas producing areas – the Caucasus, the region of Middle Asia, and the Ukrainian SSR. Other regions were to a greater extent supplied either with fuel oil (Belorussian SSR, Baltic region), being cheaper to transport (in the form of crude oil) than gas, or with cheap strip-mined coal (Kazakh SSR and the eastern regions of the Russian Soviet Federal Socialist Republic). As can be seen from Table IV, the stage

Table III. Present-Day Structure of Fuel Consumption, %[a]

Consumer category	Share in total consumption		Fuel		
			Gas	Oil	Steam coal
Total consumption	100		37	20	35
by power plants	35	100	26	35	38
by boilers	12	100	46	26	20
Small heaters	14	100	27	1	51
Industrial heaters	28	100	41	11	34
Feedstock	3	100	73	—	3

[a]Compiled from ref. [2, p. 198] with updating.

of high-rate growth of natural gas share in the boiler–furnace fuel balance lasted till the beginning of the 1970s in most regions of the country, and further increase in the gas share proceeded at more moderate rates and relatively uniformly throughout the country. Table V shows the level of penetration of natural gas in different end-use markets by the regions of the country attained by the middle of the 1970s.

Restructuring the energy use in the Soviet economy toward increased use of natural gas has enormous social and economic impacts.

The social significance of accelerated use of gas manifested itself in the improvement of working conditions, especially in industry, and in the considerable alleviation of ecological problems due to a many-fold decrease in environmental pollution resulting from the use of the cleanest fuel available. Accelerated use of gas in the domestic sector, which dramatically improved the life conditions for most of the population, had a still greater social effect. The data in Table VI show the evolution of this process. One can see, by 1975 gas supplies were available to 70% of the urban population, more than one-half of the rural population, and about two-thirds of the total population of the country. In recent years penetration of gas in domestic markets slowed down considerably due to the massive use of electricity for cooking.

Another great positive social and economic effect of growing use of gas is the considerable acceleration of technological development in energy itself and, especially, in other sectors of the economy. In particular, it greatly accelerated the introduction of many types of technological equipment, led to the use of a number of fundamentally new technological processes, enabled production of a number of new materials, etc. Unfortunately, it is not possible to estimate this effect explicitly and in detail. Nevertheless, attempts have been made in the Soviet literature to make an

Table IV. Gas Phase in Fuel Consumption by Region, %[a]

Region	1965	1970	1975
USSR	21	26	29
Russian Soviet Federal Socialist Republic	21	26	27
Ukranian SSR	25	30	31
Byelorussian SSR	17	17	16
Kazakh SSR	3	10	14
Baltic	14	16	15
Trans-Caucasus	41	35	60
Middle Asia	38	56	59

[a]Complied from ref. [4, p. 30].

Table V. Regional Gas Share in Fuel Consumption by Use, %[a]

	Power plants including		Boilers	Furnaces and industrial heaters	Feedstock
	Electricity production	Heat production			
USSR	21	26	42	30	44
Russian Soviet Federal Socialist Republic	19	24	41	29	34
Ukranian SSR	17	34	50	29	79
Byelorussian SSR	4	9	28	16	65
Kazakh SSR	12	20	14	9	–
Baltic	1	18	26	15	100
Trans-Caucasus	52	58	66	68	33
Middle Asia	66	59	65	46	100

[a]Compiled from ref. [4, p. 31].

aggregated estimate of the economic effect of creating the national gas industry [2, 6]. Since such an analysis is of both factual and methodological interest, it is useful to describe its basic logic.

To do such a retrospective analysis, it is necessary to describe, at least in general terms, a likely alternative to the massive use of natural gas. Such an alternative must have the following features:

(1) during the past period, the only practical alternative to natural gas would be coal, more precisely, coal from the Siberian basins (Kuznetsk and Kansk-Achinsk), given that the arrangements are made for its large-scale transportation to the European regions of the country;

Table VI. Gas Use in Domestic Sector, 1965–1975[a]

Indicator	1965	1970	1975
Number of gas-using apartments, million units	10.3	23.4	41.7
in urban areas	9.7	18.7	28.9
in rural areas	0.6	4.7	12.8
of which using natural gas	6.3	11.4	17.7
using liquified gas	4.0	12.0	24.0
in urban areas	3.5	7.7	11.8
in rural areas	0.5	4.3	12.2
Number of people using gas, million people	42.0	90.7	155.0
in urban areas	39.0	71.2	104.0
in rural areas	3.0	19.5	51.0
Gas share in energy use			
in urban areas	31	52	69
in rural areas	3	19	50

[a]From data in ref. [3, p. 21].

(2) in order to meet the users requirements as to the quality of energy supplies, this alternative would have to imply increased penetration of electricity in industrial and domestic sectors and also much higher use of liquid fuel (as low-sulfur fuel-oil and distillates);

(3) to lower ecological impact would be possible only through speeding up the development of district heating and cogeneration plants and, in addition through massive application of electrofilters for cleaning stack gases and the desulfurizing fuel oil.

Two fundamental presumptions (apart from a series of less important ones) underlie the analysis. Namely, both gas and non-gas alternatives of energy development should provide for: (a) the same rates and structures of economic development; (b) the same rates and directions of technological progress in energy and other sectors of the economy. These presumptions (especially the second one) are conditional and clearly idealize the non-gas alternative for the energy complex development, thereby underestimating the economic effect of the gas industry, development program.

Table VII compares the patterns of energy development corresponding to the actual gas and hypothetical non-gas alternatives. The structure of the energy complex is shown here for the starting year 1960 and in the two alternatives for 1975. In the non-gas case, as is evident from the table, it would have been necessary to provide a 320 million t.c.e. of coal extraction from 1960 to 1975, of which almost 100 million t.c.e. in the Kuznetsk

Table VII. Primary Energy Production Alternatives

Indicator	1960	1975 Actual	1975 No gas increase[a]	1975 Difference[a]
Primary energy production million t.c.e. of which	707	1600	1630	30
oil and condensate	211	696	696	0
gas	55	345	55	290
coal	369	470	790	320
Electricity production, billion kw-h	292	1037	1155	120
hydro	51	145	170	25
cogeneration		330	375	45
condensing	241	550	590	40
nuclear	–	10	20	10
Heat production, million Gcal heat-and-power		2550	2550	0
of which centralized heat production		1670	2000	300

[a]Hypothetical variant, aggregate estimate. From data in ref. [2, p. 135].

basin and 140 million t.c.e. in the Kánsk-Achinsk basin. It should be emphasized that lower efficiency of fuel utilization by end users and higher own energy needs by energy sector accompanying the use of coal instead of gas would increase the 1975 total primary energy requirement by approximately 10%.

Table VII also shows that the absence of natural gas would have required increasing the 1975 electricity production by at least 120 billion kilowatt hours (including 30 billion kilowatt hours for cooking and heating water). This leads to the growth of electric peak loads, which requires increasing the share of hydroelectric power plants (HEPP) in the overall generating capacity. Also, entailing increase in centralization of heat supply increases the need for cogeneration capacities. As a result, a relatively small amount (33%) of the additional electricity is produced by condensing power plants (CPP).

The changes in energy production technologies, naturally, would have involved drastic changes in the transportation network. In the non-gas alternative, by the year 1975 it would have been necessary to transport 275 million tons of coal from Siberian basins and the Ekibastuz deposit to

European regions of the country instead of the actual figure of 95 million tons. To provide for such a flow of coal, it would have been necessary to construct a special coal railway and to expand greatly and to reconstruct the railroad network in the European region of the country in order to redistribute the flow of eastern coals; in addition, it would have been necessary to create a coal distribution network. It is important to note here that although the average distance for transporting gas increased by 1975 by a factor of two over 1960, it is still 1.5 times less than the distance for transporting the coal needed to replace gas.

The greatest changes accompanying the non-gas alternative refer to the structure of the fuel supply to consumers. In this case the fuel supply to electric power plants (which actually consumed up to 1/3 of the gas produced by 1975) changes completely and the fuel supply for boiler plants is greatly degraded. In addition, increased electrification of the economy and a relative increase in centralized heat supply increase fuel (coal) consumption by electric power plants at the expense of boiler plants and, partly small domestic and industrial appliances. The remaining consumers should in this case switch to coal and, in part, to electricity and oil, which has to be taken almost entirely from the fuel balance of power plants.

Such a restructuring of primary energy consumption would have entailed slowing down the technological progress in power generation and probably would have greatly decreased the air quality in cities and industrial centres. Thus there would have been a considerable delay in creating and deployment of 300 MW and especially 800 and 1200 MW power units using superheated steam, large water boiler plants, etc. It is practically impossible to estimate the loss involved and it can be accounted for only qualitatively as an additional economic effect of the gas industry development program against the non-gas alternative.

Such a calculation shows that the use of gas only for industrial purposes resulted in annual savings of more than 1.7 billion rubles, which is an approximate equivalent to a saving of 6 billion rubles of capital investment and a reduction of 1 billion rubles in annual operating expenses.

Lower costs on the end-use side is only one of the components of the total economic effect of the program for developing the gas industry; the other component is lower fuel production and fuel transportation costs. For its estimation the actual cost of gas exploration and recovery was compared wtih the cost of extracting the additional amount of coal. It should be kept in mind that the technological and economic data used in the calculation correspond to the advanced coal production technology; naturally, this somewhat underestimates the cost of extracting coal in 1961–1975.

Table VIII. Estimate of Economic Effect Due to Gas Industry Development

Components of economic effect	Capital investment, billion rubles	Operating costs,[a] billion rubles/yr	Total costs, billion rubles/yr	%
Fuel production	2	1.1	1.3	19
Fuel transportation	9	1.2	3.3	47
Additional production of electricity	3	0.3	0.7	10
Fuel consumption	6	1	1.7	24
Total	20	3.5	7	100

[a]Expenditures according to the 1975 level. From data in ref. [2, p. 139].

To estimate the economic effect of fuel transportation, methods for supplying fuel to European regions of the country, the Urals, and Middle Asia in the variant without natural gas in these regions were examined. This permitted determining the amounts and directions in flows of coal needed to replace the gas. The costs of transporting coal, as in the area of its extraction, reflects promising conditions for the development of the railroad network and, evidently, underestimates the costs compared to the costs actually incurred in the 1960s.

The indicated components of the economic effect of the gas industry development program are summarized in Table VIII. This table shows that approximately one-half of the saving comes from fuel transportation. The saving in fuel production comes primarily from operating expenses. The decrease in costs due to decreased generation of electricity also made an appreciable contribution.

It is important to emphasize that the development of the gas industry freed not less than a total of 800,000 men (with respect to the 1975 level), who otherwise would have been required for extracting, transporting, and utilizing fuel. The data in Table VIII do not completely reflect the economic effect of this saving of labor resources.

The systems analysis of the large energy program required comparing its efficiency to the direct and indirect costs on its realization. Accordingly, calculations were performed using a dynamic input–output model which permitted a rough estimation of the total capital investments and the additional amounts of different forms of economic production, as well as introduction of new production capacity in allied sectors.

Comparison of the results of these calculations with the data in Table VIII indicates that the gas industry development decreased capital investments (20 billion rubles) by an amount almost equal to the capital invested directly in the industry from 1960 to 1975 (25 billion rubles). But, in so doing, about 6 billion rubles, i.e., 1/4 of direct capital investments in the gas industry, had to be invested in allied sectors of the economy. Most of these investments were spent on the development of ferrous metallurgy, since the consumption of ferrous metals for the development of gas industry over this period is comparable to their average yearly production. In addition, considerable increases in machine building (not only special, but general machinery) and construction capacity were required. All of this, undoubtedly, implied a large additional load on the economy development.

Nevertheless, a comparison of the cost of developing the gas industry with the savings produced by this industry (even with an incomplete monetary estimate of all advantages attained) doubtless indicates that this is one of the most effecient economic programs. It can be assumed roughly that each ruble of capital invested in the gas industry produced (including additional capital investments in allied sectors of the economy) more than 0.5 rubles of pure savings of capital investments with a more than threefold decrease in labor needed for fuel supply.

2. Methods for Optimizing the Gas Industry Development as a Part of the Energy Complex

The large impact of the gas industry on all aspects of energy development in the USSR permits determining correctly the potential of the industry only within an overall scheme for optimum energy forecasting. A detailed description of the methodology for energy forecasting is given in [2, pp. 18–61]. Its brief general charateristic followed by a more detailed presentation of the aspects concerning the development of the gas industry will be given below.

Long-range planning and forecasting of the complex development is based on the following initial information (see Fig. 1):

(a) A hypothesis scenario of economic development, constructed by the planning organs for a period of up to 10 years and by scientific-research organizations farther into the future;

(b) A forecast of technological progress in energy production (conver-

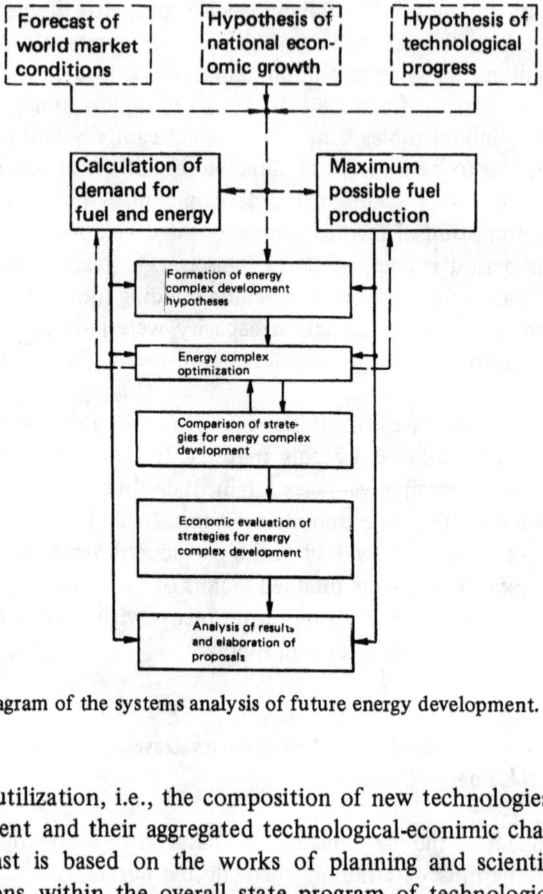

Figure 1 Diagram of the systems analysis of future energy development.

sion) and utilization, i.e., the composition of new technologies and types of equipment and their aggregated technological-econimic characteristics; this forecast is based on the works of planning and scientific-research organizations within the overall state program of technological progress and its socio-economic consequences;

(c) A forecast of the world energy development and the state of world energy markets.

Starting with this information, a forecast of the energy requirements of the economy and the possible increase in the limiting production levels of all forms of energy resources is formulated. The choice of optimal rates and periods of development of gas-bearing provinces from the marginal ones (based on geological and technological considerations) is the first stage in optimizing the gas industry development as a part of the energy complex.

Further investigation of the optimum levels and methods of utilizing gas in the future energy balance is performed in the following order (see Fig. 1).

The stage involving formulation of energy development hypotheses is intended to match previous independent forecasts of the demand for and possibilities of primary energy production. Each forecast, due to the large uncertainty in the initial information, can be obtained only in the form of a very wide range of values. The problem at this stage of investigation consists in mutual coordination of the demand for and the possible production levels of energy resources and formulation of real hypotheses of the energy complex development, i.e., determining for each variant of energy consumption (more precisely, for the corresponding scenario of economic development) the level of production of basic energy resources that can be supported by the development of the remaining sectors of the economy. An optimization input–output model, describing in a highly aggregated form the development of the energy complex and some of the most closely-related sectors of the economy, is used for this purpose. In such a model, it is possible to form a limited number of consistent (correlated according to demand and possible production levels) combinations of conditions for the energy complex development, which are optimal with respect to the criterion of minimum levelized casts and, more importantly, which take into account the actual possibilities of allied sectors; it is these combinations of conditions that form the starting hypotheses of the energy complex development.

The stage involving the energy complex optimization defines the "good" variants of the regional-technological energy structure with different combinations of external conditions, and with account of the uncertainty in the technological-economic indicators of energy facilities.

A quite detailed optimization model of energy complex, which also permits determining the system of marginal* costs on fuel and energy, is used for this purpose; if necessary, these indicators are used to adjust the economic demand of the national economy for energy resources and their possible production levels.

At the stage of comparing the strategies for energy complex development

---

*Translators's note: The term marginal cost is used here to refer to the cost of planned production output for the enterprise where the production cost is highest. In situations where certain products are interchangeable, e.g., coal and gas, marginal cost refers to the cost of the final product needed to fulfill the plan. For example, suppose gas is to be used in a new plant to be built in a region with a fixed allocation of gas. The marginal cost of the plant's output will be the cost of the more expensive fuel (coal) needed to replave the gas used by the plant. See also footnote on p. 22.

the results of multivariant optimization are analyzed in order to decrease the zone of uncertainty in the optimal solutions, for which purpose a factor analysis and methods for planning the experiment are used. In so doing, as far as possible, all available information on the conditions for invaluing all the main energy resources in the energy complex and the role of other factors examined is used.

The stage of economic evaluation of strategies allows clarifying the composition of preliminary measures that must be taken in allied sectors of the economy in order to realize the different strategies for the energy complex development. For this purpose, a special model is used to determine the dynamics of production capacities, product output, capital investments, and labor forces used in nonenergy sectors with different strategies for developing the energy complex.

The analysis stage provides a creative interpretation of the results obtained, a normative forecast of the energy complex development, and a system of arguments justifying it, as well as a formulation of the basic results of this forecast. As a result, aside from the dynamics of the energy complex development for the country as a whole and for the basic zones and regions, plans are prepared based on the following key issues:

development of each component part of the energy complex (including the gas industry), singling out the basic measures and fundamental new decisions, including the most important measures of technological progress;

Composition and content of the largest, economically significant energy programs, size and time of their implementation;

Policy in fuel and energy of conservation of consumers, including justification of the most radical changes and the marginal costs of fuel and energy stimulating them;

The most important measures on development of other sectors of the economy in order to realize the recommended strategy for the energy complex development;

In what follows, the stages in energy complex forecasting, in application to the problems of optimizing the gas industry development, are examined in succession.

2.1. *Methods for Determining Efficient Rates and Terms of Development of Gas-Bearing Provinces*

The determination of efficient regional gas-recovery dynamics presumes, first of all, a knowledge of the conditions for gas recovery, size of commercially recoverable and forecasted gas reserves in a given territory, the exploration cost, and the methods for development of fields and related costs. Second, it is necessary to know the future efficiency of gas utilization in the economy in the form of dependences of the unit efficiency of introducing additional quantities of gas into the economy on the total volume of gas used at each future time period examined. Third, it is necessary to know the monetary, material, and labor resources, which the economy can provide for developing gas-bearing provinces.

When such information is available, an efficient dynamics for developing the gas-bearing provinces can be determined by using, for example, the optimization model proposed in ref. [7], which is formulated as a linear programming problem.* The basic equations reflect the balance conditions for the production of gas reserves and for the use of gas in the economy and the limitations on the sizes of the forecasted gas reserves (by type of field) and on the alloted economic resources (primarily, capital investments). Variables of the models express the intensity of using the planned methods for preparing commercially recoverable reserves, exploiting fields, and utilizing gas as a function of time. The objective function is formulated as the maximum economic effects of gas use that are summed over the period examined and discounted in time. They are calculated for each year as a difference in economic efficiency and costs on gas production in a specific region. The objective function of the model thus corresponds to the method adopted in the USSR for economic evaluation of natural resources.[8]

It is clear that a necessary condition for using the given model is a correct determination of the resource base and the costs of preparing commercially recoverable reserves. At present the total potential gas reserves, i.e., the maximum saturation of the gas-bearing sedimentary layer, are determined from geologic criteria for the prospects of the regions based on a generalization of the most recent evaluations of geologic information, and therefore they change appreciably with time. At the same time, according to the classification adopted in the USSR, potential resources with the extracted ones include identified reserves of the discovered fields (categories $A + B + C_1 + C_2$) and undiscovered oil and gas reserves that have not yet been prospected (categories $C_2 + D_1 + D_2$). The significance of the latter for the provision of gas production increases with the expansion of the considered period of time and for a period exceeding 10 years they become the basic reserves.

*Better models and methods for optimizing the dynamics of developing gas-bearing provinces are being worked out in the USSR, but they have not yet been brought to the point of performing quantitative calculations.

To estimate the cost of transforming forecasted gas reserves into commercial reserves, a simplified mathematical model, that simulates the geological-exploratory works and realizes the probabilistic characteristics of its basic stages using the Monte Carlo method, is proposed in ref. [9]. The following basic hypotheses concerning the most important characteristics of the geologic-exploration process are used in the model:

the probability for discovering a field by exploring some promising area is not constant, and decreases with the development of the region's resources;

the average reserves of the discovered fields also do not remain constant during exploration: on the average, larger fields are discovered at earlier stages of exploration.

The following information on the results of exploratory work that had already been performed was used to derive the basic relations and to estimate the parameters of the model: (1) potential resources of the region and the distribution by horizon; (2) dynamics of exploration of promising areas by year; (3) dynamics and distribution of the average number of wells required to test a single promising area, depending on the results of the test; (4) composition of fields discovered in the region, their reserves, depths, and priority of discovery; (5) dynamics of the amount of exploratory-prospecting work and average drilling depths; (6) dynamics of capital investments in the prospecting-exploratory works with their division by type.

The following basic results can be obtained from this initial information with the help of the model simulating the process of geological prospecting:

the probability f_{ij} of obtaining an increment of reserves not less than the given value z_j with the capital investment in gas exploration k_i; by varying the expected increments to reserves ($j = 1, \ldots, l$) and the capital investment ($i = 1, \ldots, m$), a matrix of probabilities of different levels of the region development with different amounts of capital invested in exploration can be obtained;

the mathematical expectation and mean-square deviation of capital investment required in order to obtain a fixed (including maximum) increment of commercially recoverable reserves of gas in the region;

the mathematical expectation and mean-square deviation of the total increment of gas reserves with the fixed capital investment in geologic exploration;

The structure of future increments of commercial reserves of gas by fields of different size and depth.

To use the information indicated in the optimization model of the development of the gas-bearing province, the commercial gas reserves identified by simulating the process of geologic exploration are grouped by type of fields. The types are separated according to the size and depth of the fields, production rate of wells, etc. The projects of operational fields and fields being prepared for development in a given region are classified according to the same characteristics, which permits forecasting the technical-economic development indicators for fields that have not yet been discovered. In combination with the function describing the cost of transforming forecasted gas reserves into commercially recoverable reserves, as found in the simulation model of the geologic-exploratory work, this information determines, in the optimization model, the total expenditures on exploration and development of gas fields in a given gas-bearing province.

As already mentioned, in the objective function of the model for development of the gas-bearing province, the total cost of developing the province must be compared to the effect of using gas in the economy. The latter depends nonlinearly on the amounts of gas used. Such a dependence can only be obtained from multivariant optimization of the development of the energy complex by varying the amounts of gas recovered over quite a wide range by the main provinces in the country. The cost of replacing gas as it is removed from the energy balance of the country (from its substitution by nuclear power and coal for electricity and heat production to the use of electricity to replace gas in industrial processes and including losses to the economy due to a decrease in gas exports) appears as the economic efficiency of gas as a function of the volumes of gas used in the energy complex at different stages. Comparison of this function with the direct cost of developing the gas-bearing province allows the optimal dynamics of gas exploration and recovery in a given region to be determined. Efficient gas recovery dynamics differs from the optimal one by the fact that the former includes restrictions on the use of limited economic resources, primary capital, in a given region.

As shown in ref. [7], the dynamics of gas production in one of the large gas-bearing provinces in the country was investigated with the help of this technique. Investigations of the role of basic factors in the dynamics of developing a gas-bearing province, performed using the simulation and optimization models, showed the following (see ref. [10]).

1. The growth rates of commercial reserves decrease sharply in a given territory with the increase of investment in geologic-exploratory work and, in addition, the decrease is all the faster, the higher the required probability

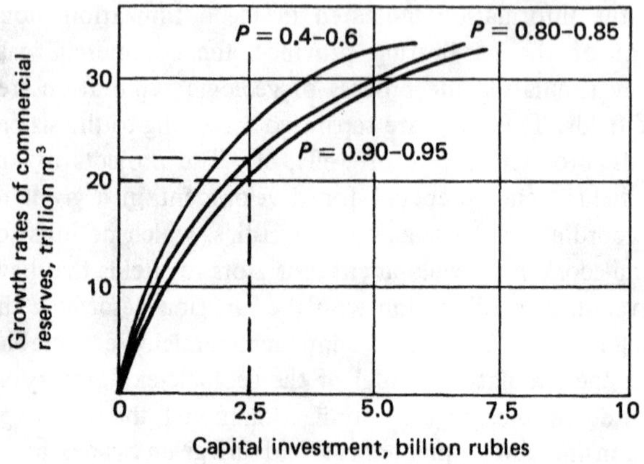

Figure 2 Growth rates of commercial gas reserves as a function of capital investment in exploration. *P* is the probability of growth rates in commercial reserves.

P of their production (see Fig. 2). As a result, for the acceptable costs (see the fine dashed lines in Fig. 2), with high probability (exceeding 0.8), it is possible to count on transforming only approximately one-half of the forecasted gas reserves into commercially recoverable ones.

2. Multivariant calculations have established the following property of the change in the production cost in the process of region development (see Fig. 3): the maximum unit cost of gas recovery (i.e., the costs on the worst fields being developed at a given stage) remains practically constant during the entire period of fuel production growth in the region and begins to increase, with growing intensity, only when the region enters the decreasing production phase.

3. Different hypotheses about the sizes of the forecasted gas reserves in a region (or, analogously, different probabilities of obtaining commercially recoverable reserves, see Figs. 2 and 3) differ little at the initial stage of the region development but appreciably affect the maximum production level and rates of production decrease.

4. Limitations on capital investments have the greatest effect on the dynamics of developing a gas-bearing province: their decrease as compared to the capital investment on the optimum scheme region development causes a relatively larger decrease in the increment to reserves and to production levels, postpones attainment of maximum production, and can even decrease its size, but slows down the rates of subsequent decrease in production and extends the time for developing the gas-bearing province.

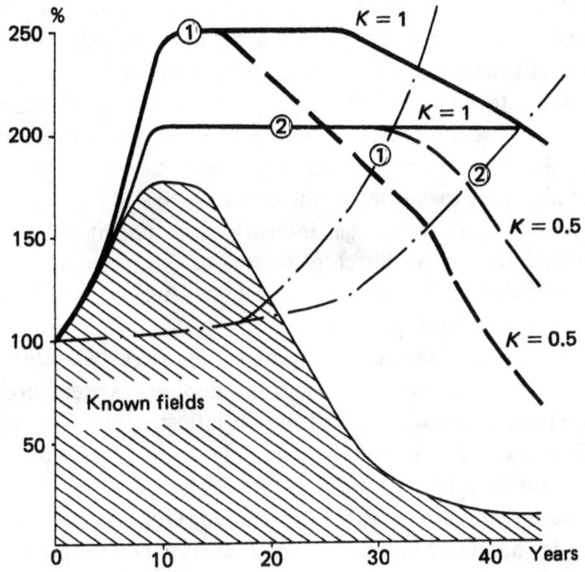

Figure 3 Dynamics of developing a large gas province. ——, – – –, annual gas production; *K*, coefficient of confirmation of forecasted gas reserves; —·—, ultimate cost of gas production with $K = 0.5$; (1), high production level; (2), low production level.

The examined technique for determining the efficient dynamics of developing gas-bearing provinces requires the use of information (economic efficiency function for gas utilization, sizes of capital investment in gas exploration and production) that can be obtained only by a quite detailed investigation of the prospects for the energy complex development, i.e., at later stages of its optimization (see Fig. 1). For this reason, efficient gas production levels by stages of the studied period can be obtained only from a certain iterative procedure, which successively refines the concepts of the prospects for developing the gas industry in the framework of the energy complex.

2.2. *Methods for Optimizing Production and Distribution of Gas within the Energy Complex*

It was seen that the summation of efficient dynamics of developing gas-bearing provinces gives the total gas production levels by stages of the studied period, i.e., it solves the main problem of determining the future development of the gas industry. In reality, the gas-production dynamics obtained in this manner, can be viewed only as upper limits of production levels (so-called technically feasible production) due to the simplification of the methods used. These estimations must be made more precise (in practice, in the direction of their decrease) owing to more detailed account of, first of all, real capabilities of the economy to share out money, materials, and labor forces for the gas industry development and, second, the actual demand for gas by different sectors of the economy. Both points have been included in the scheme for systems studies of the prospects for the energy complex development (Fig. 1).

The next (second) stage in this scheme is the aggregated optimization of the energy and allied sectors: ferrous and nonferrous metallurgy, several machine-building sectors, the construction industry, etc. (for more detail see ref. [2, p. 30]). This permits determining the real ranges of production of the basic types of fuel by stages of the studied period (within the limits found above on technically feasible production) for two or three scenarios of economic development and related energy consumption scenarios [2, 11].

In such an optimization input–output dynamic model, the gas industry is represented by the main gas-bearing provinces. The interchangeability of gas with coal, nuclear power, and oil (displacement of fuel oil from power plants and replacement of oil exports by gas exports) is taken into account.

The solution, obtained using the input–output model, by the criterion of minimum total discounted costs developing the energy and allied sectors, refines the dynamics of natural-gas production growth in accordance with the requirements and capabilities of the economy within the limits of the technological possibilities and taking into account the effecient development of the remaining sectors of the energy complex. But, together with such a determinate trajectory, it is no less important to estimate the real ranges of variation in gas-production levels and, therefore, of energy resources competing with gas, primarily, coal. For this purpose, using the same model, optimization was performed according to two criteria: (1) maximum natural gas share (therefore, minimum coal) and (2) maximum coal share, which automatically corresponds to minimizing the gas share. The range of possible variations in the share of natural gas and that of coal in the energy balance of the country, obtained under these conditions, were determined. This range has the following characteristics (see Fig. 4):

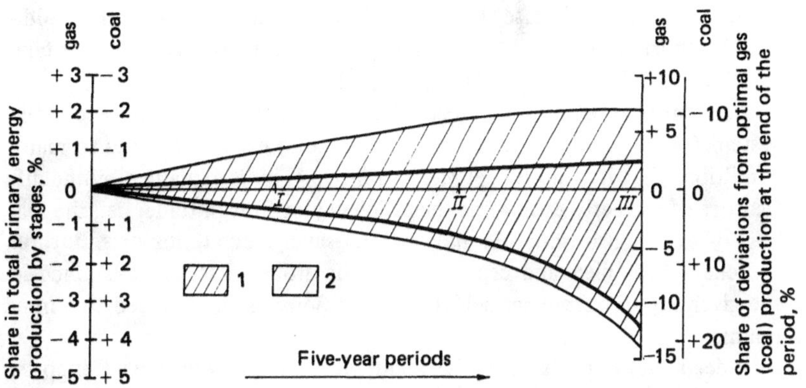

Figure 4 Gas and coal substitution potential in the USSR energy balance. (1) internal possibilities. (2) Same with account of equipment imports

(a) The possibilities for changing the contribution of the indicated forms of fuel are relatively small at the initial stages of the studied period, but increase considerably with time;

(b) Toward the end of the studied period the possibility for changing the structure of developing the coal and gas industries increases to 3–4% of the total production of energy resources. This corresponds to a 10–12% change in the amount of coal and gas recovered, i.e., several tens of millions of tons of coal equivalent. Nevertheless, the limitations arising from the allied sectors on the possiblilities of structural changes in the energy complex must be recognized as being very great;

(c) Attention should be paid to the asymmetry of the identified domain with respect to the curve of the optimum structure of energy complex. This curve clearly approaches the upper limit of possibilities for using gas and, therefore, to the lower limit for coal extraction. It follows that for the (very rough, especially in the part of accounting the territorial factor and the effect for consumers), description of interchangeablility of coal and gas used in the given model, the use of gas has definite economic advantages.

The results described above correspond to the energy complex development based on domestic production of the required equipment, material, etc. (more precisely, on production with account of fixed imports). In reality the possibilities for altering the gas and coal shares can be additionally increased by increasing imports of the required equipment and materials. But, in this case, there arises the questions: what level of additional imports can we talk about and, more importantly, what does it take to achieve this level of imports?

Account of foreign trade outside the framework of the overall economic system (i.e., in studying some group of sectors) usually involves so many conditions that it is no longer valid to do so. This is explained by the difficulties of correctly estimating the efficiency of foreign trade, i.e., the real cost of exported production and, mainly, savings due to imports. But, for the energy complex and especially for the group of sectors examined in the given input–output model, these conditions can be reduced to a minimum.

Indeed fuel exports comprise about 40% of the total Soviet exports, when the produciton of ferrous metallurgy and machine-building sectors examined in the input–output model are included, this share increases up to almost 50% [12].

Taking into account the conditions of production of such a large fraction of the exports permits estimating quite correctly the economy costs related to some (relatively small) changes in the total volume of exports.

The situation is different when taking into account the efficiency of imports in the group of sectors under study, since most of the imports fall into other sectors of the economy. But as soon as we talk about relatively small changes in foreign trade, it is important to know for the investigations not the total, but the unit savings resulting from imports. The latter must be determined not on the average, but by the marginal forms of imports, since it is these imports that are purchased using the money earned from the additional exports.* As for the composition of the marginal imports, a very plausible hypothesis is that such imports also include production used on the energy complex development, in particular, pipes and compressors for gas pipelines.

If this hypotheses is adopted, then certain adjustments of the export–import links of the energy complex within the framework of the input–output model used are methodologically justified and useful. The efficient sizes of imports of non-production for developing the energy complex can then be determined from the following considerations.

The composition of imports is not fixed *a priori*; it is determined in the

*Marginal is understood here to mean the forms of production that must be purchased last with optimum disrtibtion (according to the criterion of maximum efficiency) of limited import possibilities.

model starting from the requirements for eliminating bottlenecks in the energy complex development. On the other hand, the volume of additional imports is determined by two circumstances: (a) it is completely compensated by the increase in fuel exports and (b) the maximum effect is achieved for the complex of sectors examined.

In the calculations performed, the fuel exports did not include oil and nuclear fuel. The amounts of the remaining forms of fuel and electricity exported were determined as a result of optimization; for the adopted ratios of world prices, natural gas completely provides the increase in exports.

The method used to determine the most efficient volume of imports is conveniently explained in Fig. 5. Here the volume of imports of equipment required (V^i) is laid on the horizontal axis and the profit from gas exports obtained due to imports of equipment (P^e) is shown on the vertical axis. For a number of reasons (decreased efficiency of imports after eliminating the worst bottlenecks, transition to the worse conditions for recovery of additional gas, etc), as the imports V^i increase, the profit P^e, as a rule, at first increases rapidly and then more slowly. Moreover, it strongly depends on the prices of the imported equipment and, especially, the prices of the exported fuel (see curves for minimum (C_{min}), probable ($\bar{C}$), and maximum (C_{max}) gas prices in Fig. 5).

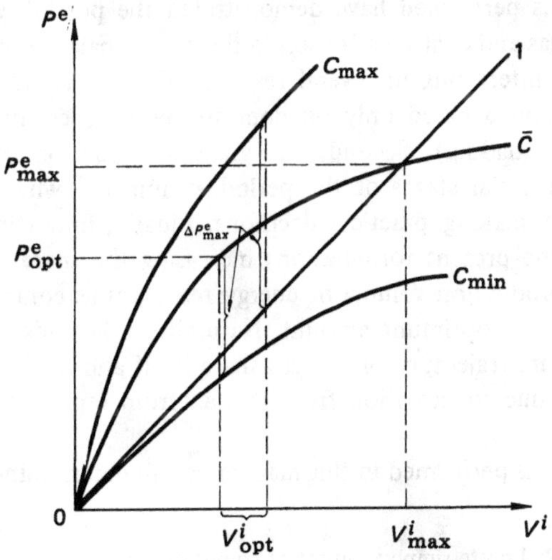

Figure 5 Illustration for the determination of the desirable level of energy equipment imports.

It is clear that importing equipment under the assumptions made is justified only if the profit obtained in this case exceeds the cost of the additional fuel exports, i.e., the corresponding curve at least in some range of V^i lies above the line 0–1, drawn at an angle of 45°. Taking this into account, it is not expedient to import equipment with the minimum gas prices within the framework of the approach examined. For the probable gas prices ($\bar{C}$), on the other hand, imports are justified within limits up to $V^i_{\mathbf{max}}$ (see Fig. 5). But, if the volume of imports is taken as equal to $V^i_{\mathbf{max}}$, then the entire volume of the additional gas recovered must be exported to compensate for the imported gas equipment, while domestic consumers will not obtain any of the additional gas. This is completely unjustified and it is necessary to aim for a level of equipment imports ($V^i_{\mathbf{opt}}$), for which the maximum additional profit, obtained by the economy after compensating for expenditures on imports, is achieved:

$$\Delta P^e = P^e - V^i \rightarrow \max . \qquad (1)$$

As shown in Fig. 5, the volume $V^i_{\mathbf{opt}}$ depends on the conditions (primarily, on prices in the foreign market), but varies over a quite narrow range. Its values under different conditions are determined directly by calculation using the optimization input–output model. The additional amount of fuel, obtained due to equipment imports, is determined automatically at the same time.

Calculations performed have demonstrated the possiblity of enlarging the range of gas and coal substitution in the energy balance of the country. This range is interesting in several respects. First, it is much wider than the range obtained based only on domestic possiblities (noted in Fig. 4 by the close shading). Second, the greatest expansion of possibilities occurs at the initial stages of the period examined,* which is especially important for making practical decisions. Finally, importing equipment requires (in the present formulation) increasing the share of natural gas in the total production volume of energy resources as compared with the previously found optimum amount. Naturally, this leads to the change in the optimum trajectory of the gas share itself and in the absolute size of the losses due to deviation from it, i.e., from replacement of gas by coal.

Investigations performed in this manner permit determining for different

*This is explained quite simply: import of equipment is not inertial, i.e. it can provide an increment to fuel production with a delay equal only to the time required to install this equipment.

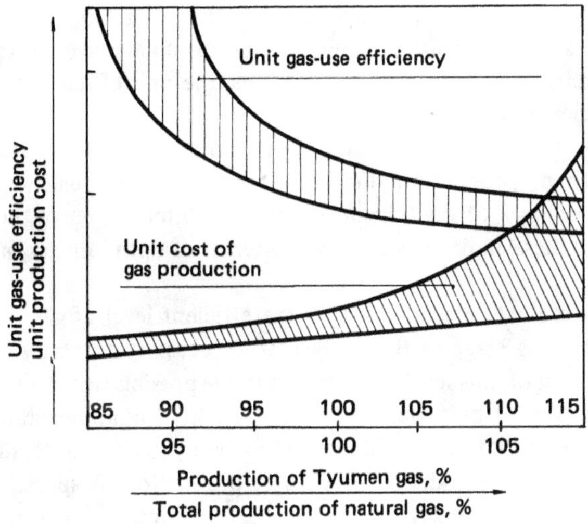

Figure 6 Efficiency of natural gas use in the energy complex.

scenarios of economic development the corresponding possible ranges of natural gas utilization in the energy complex. The ranges of development of the primary fuel bases are identified at the same time.

Multivariant optimization of gas distribution over territories and categories of consumers is performed within the gas production limits found in the input–output model. A detailed optimization model of the energy complex, described in detail in refs [2] and [13], is used for this purpose. In this model, optimum gas production levels in different regions of the country, basic interregional gas flows, and gas distribution over 6–8 categories of consumers in each region (depending on the duration of the studied period, the model considers from 12 to 26 regions) are determined in combination with other sectors of the energy complex. Such a distribution of different levels of total gas production by territory and by consumer gives the most correct determination of both the costs of gas production and transportation and the gas utilization efficiency, which permits determining the efficient future development of the gas industry as well.

The results of such an investigation are shown in Fig. 6. Here the amounts of natural gas produced in the country as a whole as a percentage of the maximum production level, found at the preceding stage of the analysis, i.e., while studying the input–output model, are shown on the horizontal axis. The specific indicators of the utilization efficiency and expenditures on gas production and transportation are given in the vertical axis. Naturally, as gas production increases, the unit efficiency of gas utilization in the energy complex has a tendency to decrease. On the contrary, the unit cost of gas production and transportation will increase progressively with growth of production due to the use of lower quality fields and increased distances of gas transportation. The intersection of the domains of values of these indicators gives the interval of optimum gas production levels.

To make a final determination of the efficient level of gas production, at the concluding stage of the analysis it is necessary to verify once again the realizability of this level with respect to the possibilities of allied sectors. However, this must be done not in the aggregated nonomenclature of the branches of (metallurgy, machine building, etc.), but by a detailed set of many tens or even hundreds of types of production. A special model of the inverse impact of energy sector on the development of related sectors was constructed for these purposes [2, 14, 15]. The model is intended for making comparative evaluations of different variants of developing the energy complex as a whole, its individual sectors (in particular, gas industry), or objects. Calculations using this model give the dynamics of the necessary additional output of different forms of production both for the energy and allied sectors, whose productivity changes appreciably in different variants of the energy complex development. In addition, the model determines the dynamics of adding production capacities required by the production output dynamics found.

The investigations showed that a change in fuel (especially gas) production often affects the production volume of those sectors whose production is not directly consumed in the energy complex, and even makes it necessary to add to the production capacity of these sectors. At the same time, the lead time for introduction of new capacities increases in length with the level of relations between the given sector and the energy complex i.e., the longer the technological chain of relations between the given sector and the energy complex the longer is the lead time. As a result, for most related sectors, production capacities are added beginning from the preceding five years and it is here that the maximum introduction of capacity occurs.

Analysis of the necessary time periods for increasing production capacity allows quite objective judgement on the realizablility of the strategies being compared: if these capacities are already required in the first years of the studied period, while the reserve capacities of this form are absent in the economy, then it is necessary either to reject the strategy examined

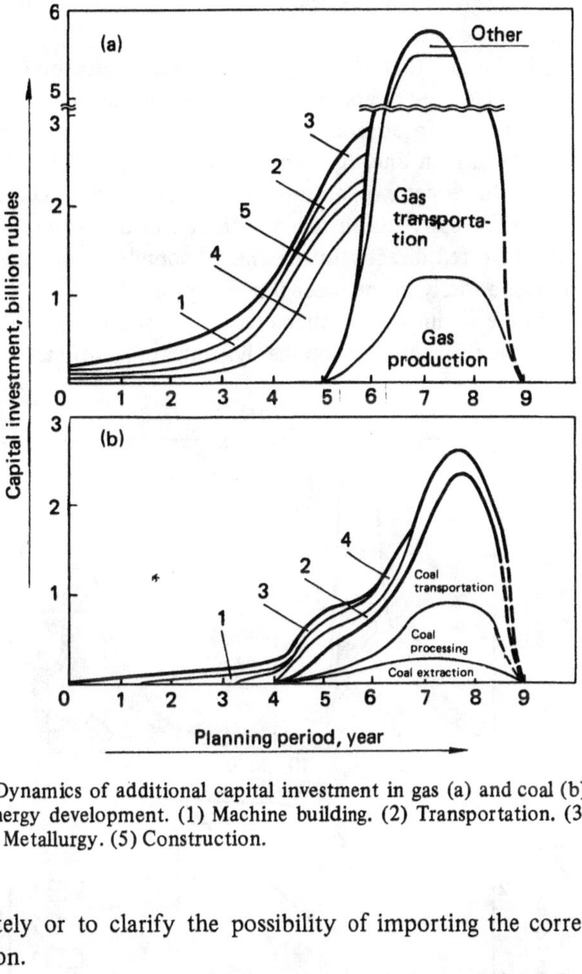

Figure 7 Dynamics of additional capital investment in gas (a) and coal (b) strategies for the energy development. (1) Machine building. (2) Transportation. (3) Miscellaneous. (4) Metallurgy. (5) Construction.

immediately or to clarify the possibility of importing the corresponding production.

The dynamics of additional capital investment in the development of the energy complex and the sectors of the economy supporting it is determined for the remaining strategies (checked for realizability). As an example, this is shown in Fig. 7 for the strategies of energy complex development differing only in the increased gas production in one (Fig. 7*a*), and the increased coal extraction by the same amount in the other at the end of the studied period (Fig. 7*b*).

The thick lines in Fig. 7 show the dynamics of direct capital investment in these measures (with their division by cost components) and total capital investments, while the thin lines show the required capital investment in other sectors of the economy.

It should be noted that the preparatory capital investment in other sectors must be made ten years in advance (even for normative periods of constructing objects). In addition, six to seven years before the energy facilities are brought on line, the capital investments are already so large that they equal the direct expenditures on fuel production. On the whole, however, capital investments in allied sectors constitute here about 1/4 of the capital invested directly in the energy complex, but they are concentrated almost entirely in the preceding five years.

Labor costs are an important characteristic of strategies for developing the energy complex. Figure 8 shows the dynamics of additional labor forces

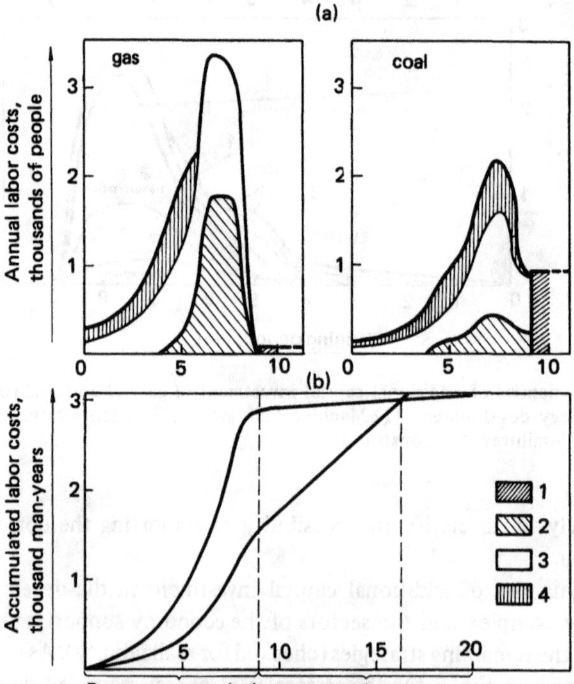

Figure 8 Dynamics of labor costs on increasing fuel production (a) Dynamics of direct and indirect annual labor costs; (b) accumulated direct and indirect labor costs. Direct labor costs: (1) exploitation; (2) construction. Indirect labor costs: (3) exploitation in related sectors; (4) construction in related sectors.

used for exploitation and construction in the gas and coal industries, and as a result the related sectors of the economy. As is evident from Fig. 8*a*, the annual labor costs in the economy, associated with additional utilization of gas during the construction period, are approximately 1.5 times greater than labor costs of increasing the use of coal. However, in subsequent exploitation they are naturally much lower for gas alternative than for the coal alternative. Thus, as shown in Fig. 8*b*, the large integral labor costs of construction of enterprises for gas recovery and transport are compensated by their savings from exploitation within only five to six years.

The examples presented (see Figs. 7 and 8) show the considerable difference in economic consequences of realizing equal increments of production of different fuels. Taking this into account, in cases of approximately identical economic efficiency of developing the gas and competing industries within the energy complex preference must be given to the less capital intensive industries.

2.3. Methods for Making Complex Evaluation of Large Programs for the Gas Industry Development as a Part of the Energy Complex

Rapid rates of the gas industry development and a large share of gas in the overall production of energy resources require economic evaluations, within the energy complex, both the total gas production levels and the main programs for the gas industry development. These include, first of all, programs for complex development of large oil-bearing provinces and, second, programs of technological progress in the gas industry, primarily, gas transportation. All of these factors strongly influence the overall dimensions of gas utilization in the economy and also require very large expenditures. For this reason, their objective evaluation and comparison with alternative programs can be obtained only within the energy complex. It is useful to illustrate the methods used to make such evaluations on the program of developing the Western Siberian oil–gas-bearing province [2] and on the program of technical re-equipment of main gas pipelines [16].

Development of the Wester Siberian oil–gas-bearing province, reaching an economically justifiable level of oil and gas production, is one of the largest economic programs in the USSR. This program includes the following:

exploration and recovery of the required volumes of oil and gas;

their distribution with the help of large pipeline systems running to the Far East and Western Europe;

expansion of the industry on oil and casing-head gas refining and gas-condensate recovery, as well as petrochemical complexes based on the raw material obtained;

additional development of metallurgy and machine building to supply pipes, equipment and materials for the oil, oil refining and petrochemical industries, and to create the necessary infrastructure in poorly developed regions of Western Siberia.

Efficient levels of oil and gas production in Western Siberia are determined by the multivariant optimization and the factor analysis of the conditions for the energy complex development. As is already mentioned, the economically justifiable oil production level in the country will to a large extent be provided by the additional recovery of oil in Western Siberia. An analogous situation also exists for the recovery of the Tyumen natural gas, whose efficient production level will greatly exceed the present gas recovery level in the country.

Based on optimization calculations, it is possible to estimate the total economic efficiency of the program for developing the Western Siberian oil-gas-bearing province.

The alternative hypotheses for developing the energy complex gives rise here to the greatest difficulties. With a large degree of conditionality it may be assumed that in the absence of the Western Siberian oil–gas, bearing province the oil and gas industry in the country would develop on the basis of oil-bearing regions (primarily in Eastern Siberia) that have been recently discovered and are at the exploration stage. In addition, in the alternative variant of the energy complex, it is necessary to provide for large-scale introduction of Siberian coals, as well as more rapid development of nuclear power within the limits determined primarily by the increase in the electrification level of the economy with decreasing use of high-grade fuel.

The main component of the efficiency of the Western Siberian province is the savings at the expense of extraction and transport of Siberian coals: Kuznetsk and Kansk-Achinsk coals. This is explained both by the considerable difference between the corresponding unit costs (7–9 rubles/t.c.e.) and by the large volume of required additional coal. This component is closely related to savings from the use of high-grade fuel by consumers. Approximately one-half of the savings is achieved at power plants, but the savings in other consumer categories, primarily boiler plants, are also important.

Each of the remaining components of the savings does not have a decisive significance, but in sum they constitute a large amount. They include savings from substitution of more expensive oil and gas resources, savings from oil refining (their redistribution between primary and secondary processes), savings from decreasing the required nuclear power plant capacity, as well as the effect of oil and gas exports.

On the whole, the Western Siberian oil–gas-bearing province, according to our calculations, is capable of giving, even during the first period of its exploitation, a pure saving, exceeding all expenses incurred, including creation of the regional infrastructure.

Refining the recovered oil and gas serves as an important integral part of the program for developing the Western Siberian oil–gas-bearing province. This includes refining oil and casing-head gas (in gas refining plants) and "drying" gas from gas-condensate fields.

Calculations have shown that using casing-head gas as a petrochemical raw material and shift of the production of raw material for petrochemistry to Siberia can result in a large economic effect: the annual economic effect at the stage of complete development, according to our estimates, can reach one billion rubles; in addition, the conditions for supplying fuel to small users in Siberia are improved considerably due to the larger resources of petroleum fuel here.

Delivery of the recovered oil and gas to users is the most complicated part of the program for developing the Western Siberian oil–gas-bearing province. Optimization of the energy complex shows that most of the oil and, especially, natural gas must be supplied beyond the boundaries of Siberia: to consumers in the European regions of the country and for export. For this purpose, it will be necessary to construct every year large-diameter pipelines with a length of 2.5–3 thousand kilometers including a complete set of compressor stations and an appropriate distribution network. It will also be necessary to take additional measures for transportation of exported oil and gas, in particular, extending gas and oil pipelines into European countries. Realization of these measures will require large expenditures of material especially in connection with the necessity of laying large sections of pipelines under permafrost conditions, this fact greatly increases the cost of construction and requires solving complicated technical problems.

The measures required by the program examined are as follows: (a) completion of the required exploratory work and (b) timely development of production capacities in metallurgy, rolling, machine building, and construction industry.

The planned volumes of exploratory work, oil and gas production in

Table IX. Approximate Estimation of Demand for Industrial Production to Develop the Western-Siberian Oil-Gas-Bearing Province[a]

Type of production	Demand for the period	Period	
		First five years	Second five years
Ferrous metallurgy:			
Pig iron, million tons	125	50	45
Steel, million tons	210	80	75
Rolled metal, million tons	175	65	60
Pipes, million tons	130	50	45
Construction materials:			
Cement, million tons	75	35	30
Precast reinforced concrete, million m^3	100	5	40
Specialized machine building:			
Drilling equipment, thousand units	3.7	2.1	1.1
Gas pumping units, million kw	25	10	10
Oil pumping units, thousand units	1.5	0.9	0.5
General machine building:			
Construction machines and mechanisms, thousand units	200	100	80
Rolling stock, thousand units	100	50	40
Machine-tools and forging equipment, thousand units	95	60	30

[a]Compiled from data in ref. [2, p. 214].

Western Siberia, and their transportation by pipelines to the European regions were estimated from the point of view of the required expenditures of different forms of economic output. For this purpose, the direct and indirect expenditures of the basic types of output on the production and distribution of Western Siberian energy resources and the production capacities required to produce this output were determined for the period examined with the help of the dynamic input–output model. Table IX shows some results of this calculation. These results take only partial account of expenditures on creating an infrastructure and exclude expenditures on petroleum refining, petrochemistry, and gas refining plants. But even with such assumptions, the program for developing the Western Siberian oil–gas-bearing province requires very large expenditures of material in the period examined. If they are compared with the current level of output of the corresponding production, then more than one-half of the annual output of construction materials in the country, 1.5 times more pig iron, steel, and rolled iron are required to realize the program. Extensive development of the oil and gas industry can be provided only by the timely creation of large additional production capacities in other related sectors as well: plants for producing compressors, pumps, and electric motors, cement plants, etc.

Efficient paths for meeting this demand require a special study. The most complicated problem is the relation between the import and domestic output of the required production. This problem cannot be solved by direct economic calculation due to the uncertainty of world market conditions, etc. Neglecting these factors, it is necessary to import large-diameter pipes, deep-drilling equipment, and some other types of production to realize the planned program in time. Otherwise it would already be necessary in the near term to have free capacity in ferrous metallurgy, equivalent to a large metallurgical integrated plant.

According to model calculations, the total capital investment required to realize the program is estimated at 65–70 billion rubles. In addition, the development of allied sectors (metallurgy, electrical power industry, construction industry, machine building, railroad transportation) will apparently require 20–25 billion rubles. These investments will not only be paid for in the normative (eight year) period, but will provide a saving of about 50% of all expenditures on oil and gas production from Western Siberia. These large additional savings ("extra profit") permit viewing the program for developing the Western Siberian oil–gas-bearing province as one of the most effective economic programs in the period under study.

The methods used to have systems evaluation of the important directions for technological progress in the gas industry can be conveniently explained on an example of justifying the expendience of and choice of parameters of low-temperature gas transportation. The need for examining this problem within the energy complex stems from the fact that due to the sharp decrease in the use of metal and pipes for gas pipelines the corresponding measures allow great increase of the natural gas reserves contributed to the energy balance and great changes in the overall requirements of the energy complex for development of the related sectors of the economy.

Multivariant optimization of main gas pipelines with the usual parameters for gas transportation, with the gas cooled to 210 K and in the liquified form (temperature 150–160° C) established the relations, presented in Table X, for the main technical-economic indicators for gas transportation (for more detail see ref. [16]). It is evident from them, in particular, that

Table X. Cost of 3000 km Gas Pipelines[a]

Pipeline diameter (mm)	Transportation mode	Capacity (billion m^3/yr)	Unit costs on 1 km billion m^3/yr		
			Metal Consumption (ton)	Compressor activity (kw)	Capital investment (thousand rubles)
1220	Usual	18	25.6	15.3	14.6
	Cooled	40	16.0	16.6	13.3
	Liquified	70	8.1	17.2	9.0
1420	Usual	30	25.0	14.1	13.2
	Cooled	60	14.3	13.8	11.3
	Liquified	90	7.5	13.9	8.4

[a]Compiled from data in ref. [16, p. 37].

the metal use for gas transportation decreases by a factor of 1.6 for transportation of cooled gas and by a factor of 3.2 for transportation of liquified gas. This means that with the same capabilities of the economy for sharing out metal for construction of pipelines, it is possible to greatly increase the volume of gas transported and, therefore, of gas used and to obtain, as a result, a large additional economic effect.

The determination of the value of this effect and the adjustment, considering this effect, of the efficient ranges and dimensions of applying different methods of gas transportation require that they be compared within the framework of energy complex optimization. The corresponding calculations showed that rejection of low-temperature gas transportation decreases the volume of gas utilized in a number of regions of the country and thereby sharply degrades the conditions for supplying fuel to electric power plants and some boilers. As a result, it becomes necessary to additionally construct nuclear power plants and to extensively reconstruct mines in the European coal basins with extraction costs equal to 32–35 rubles t.c.e. For the sake of eliminating the related losses it is useful to considerably expand the application of low-temperature gas transportation as compared to that determined by the usual technical-economic calculations.

To choose the correct method for gas transportation, it is important

Table XI. Change in the Economy Costs on Low-Temperature Gas Transportation Against the Usual Form[a]

Indicators	Cooled gas	Liquified gas
Capital investment in gas transportation, million rubles	− 50	− 1500
Capital investment in related sectors, million rubles	− 450	− 820
Consumption: large-diameter pipes, million tons	− 3.5	− 7.5
Rolled metal, million tons	− 3.4	− 6.8
Cement, million tons	− 1.6	− 2.6
Chemical production, million rubles	+ 190	+ 450
Gas pumping units, million kw	− 2.7	− 4.5
Chemical and pumping–compressor equipment, million rubles	+ 760	+ 1750
Construction machinery and equipment, million rubles	− 105	− 182
Metallurgical equipment, million rubles	− 70	− 125

[a]From data in ref. [16, p. 38]; + indicates an increase and − indicates a decrease compared to the usual technology.

to estimate the requirements for the development of related sectors of the economy in connection with the use of new transportation technologies. To illustrate this, Table XI shows the changes in capital investment and basic types of material expenditures when the usual gas transportation is replaced by a low-temperature one. As is evident, this requires a considerable growth in production of refrigeration and pumping-compressor equipment, but it gives a large saving in pipes, rolled iron, and cement and decreases the total capital investment in gas transportation and allied sectors.

The considered methods for economic evaluation of large programs for developing the gas industry and for its technological progress assure a better choice of the development rates and the structure of this most important industry of the energy sector at the new stage.

3. The Role of Natural Gas in the New Energy Strategy of the USSR

3.1. New Conditions and Concepts for the USSR Energy Development

During the last 20 years, the rapid growth of oil and gas production (see Table I) has occurred with comparitively low economic expenditures. This was ensured to a large extent by concentrating on the development of the the most effective fields in very large oil–gas-bearing provinces. This policy was undoubtedly correct at the initial stages of gas industry development and it is completely justified during the period of forming a large-scale oil production industry. However, in terms of the long-range future, gradual depletion of the largest and most favorably situated fields requires that these fields be compensated and especially that the increases in demand be met by smaller, less favorably located fields with low production rates. As a result, the economic costs on producing and transporting oil and gas have a distinct tendency to increase. For oil, the growth of costs is conditioned primarily by the increase of well depths and by the reduction of averate production rate of wells. For natural gas, these parameters were only improved, but the average distance of gas transportation increased rapidly [17]. As a result, the unit capital investment in recovery, transportation, and refining of oil and gas increased steadily both per unit of newly introduced capacity and, especially, per unit increase of oil and gas production. The latter is due to the increasing removal of operating wells from exploitation. As a result, the unit capital investment to increase fuel production began to exceed the capital investment per unit capacity by more than a factor of three for oil and by 25–30% for gas.

The above factors responsible for the increase in the cost of oil and gas, will become much more significant in the future, so that the capital investment per unit of new capacity will increase by more than a factor of 2.5 for oil and by more than a factor of 2 for gas (see Fig. 9).

The new conditions for energy development will in many ways change the concept of the energy balance in the foreseeable future. They differ to such an extent from the conditions operating at the preceding stages that they essentially require the formation of a new energy strategy. The essence of this new strategy consists in *completion of the multi-year of increase in the share of hydrocarbon fuel – oil and gas – in the overall production of primary energy, and in gradual transitin to more abundant energy resources.* The main aspects of the new energy strategy for the productive structure of the energy complex are as follows (see Fig. 9) [18].

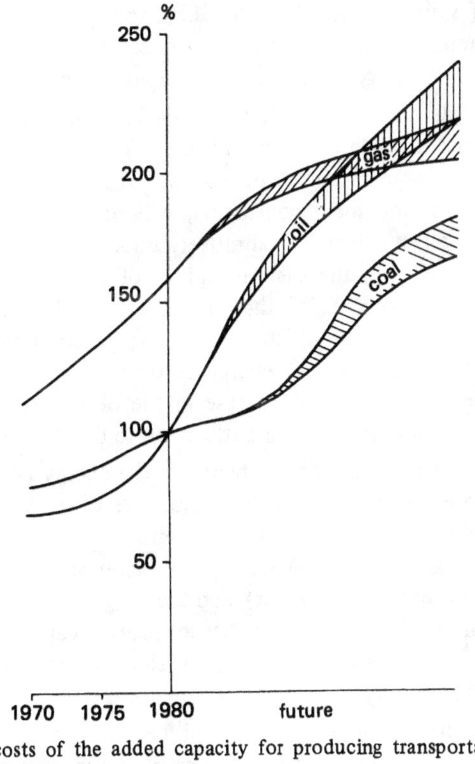

Figure 9 Unit costs of the added capacity for producing transportation, and conversion of basic fuel forms.

1. The growth rate of the oil industry, which was responsible for most of the main increase in the production of energy resources during the last 20 years, slows down under the new conditions. For this reason, while making every effort to continue to increase oil production (at first by increasing the amount of drilling and by developing new fields and later by using new methods for extracting oil from beds and increasing the development of new off-shore oil-bearing fields), emphasis will have to be placed on its conservation and substitution by less expensive and more abundant energy resources. The most important and urgent measure in this respect is to decrease as rapidly as possible the use of oil as a boiler-furnace fuel (especially by power plants) and to drastically increase secondary oil refining capacity (for production of white petroleum products). Together with accelerated shift to diesel fuel use in transport, the use of compressed gas and electrification of transport, this can significantly slow down the rate of growth in the demand for liquid fuel. Large-scale production of synthetic liquid fueld (SLF) from gas, coal, and shale will create the conditions for a painless decrease in oil production in the future, if this is necessary to secure the required resources.

2. Natural gas must play a decisive role in replacing oil as a boiler–furnace fuel and to a large extent in exports of energy resources. The presence of very large resources of natural gas in the country, although far removed from the main consumers, will not only provide for the replacement of oil, but it will qualitatively improve the energy supply to consumers, especially in the eastern regions of the country. In this connection, rapid development of the gas industry will become a primary factor in the development of the USSR energy complex. In addition, its share in the production of energy resources must increase to 35% (completely compensating the decrease in the oil share) and, after some period of stabilization, it will apparently begin to decrease (see Fig. 10).

3. The second most important factor in the energy complex development in the foreseeable future is the utmost use of nuclear power. In the European regions of the country, nuclear power plants (NPP) must provide the entire increase in the base load (including a significant part of the replacement of obsolete equipment), and then together with nuclear boiler plants most of the increase in the district heating capacity. In this area, nuclear power will replace natural gas, thereby stabilizing the gas production levels in the country.

4. In the forthcoming period, there will be an urgent need to accelerate the development of the coal industry (one of the most inertial sectors of

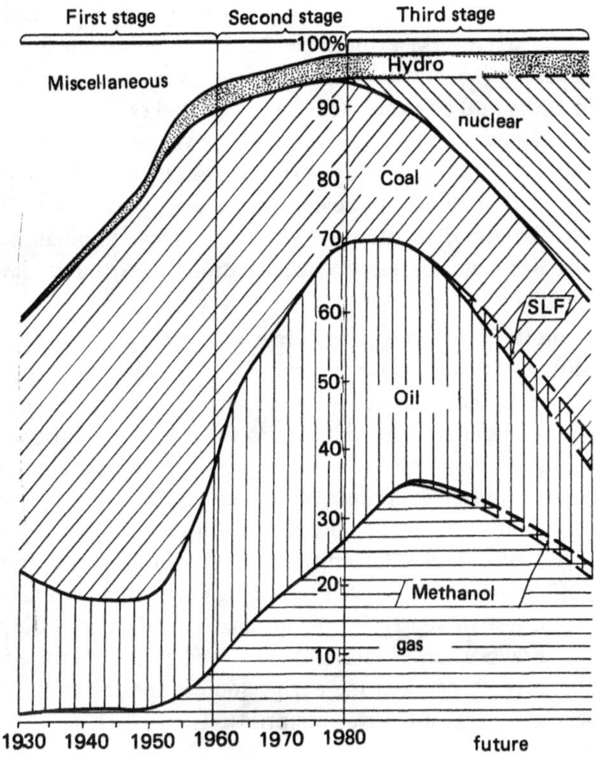

Figure 10 Structure of primary energy production in the USSR.

the energy complex). The necessary large shift in its development will require large-scale development of strip mining of coal in the eastern basins of the country (Kuzetsk, Ekibastuz, and Kansk-Achinsk) with the coal used primarily as a fuel for generating electricity and producing SLF.

5. In the period under study, favorable conditions are created for wider use of hydro-power resources (including pumped storage plants for regulating the electric load curve, providing more favorable conditions for operation of nuclear power plants) and nontraditional, especially renewable, energy sources (solar, wind, geothermal, biomass, etc). Nevertheless, in the foreseeable future, their fraction in the overall production of energy resources, even including hydro power, will apparently not exceed 5% (see Fig. 10).

6. At the stage under study, the technological progress in energy sector,

together with an increase in production efficiency, must proceed so as to expand the resource base as much as possible. About one-fourth of the total production (i.e., up to one-half of the increase in production) of energy resources in the forthcoming 20 year period can be secured by such fundamentally new technologies as nuclear power, production of artificial fuel, tertiary oil recovery, development of the Arctic shelf, d.c. transmission lines in order to bring eastern coals into the energy balance of European regions, etc. An important step will have to be taken to obtain secondary nuclear fuel (fast-neutron reactors, regeneration cycles) and to develop the fusion technology.

7. Transportation of energy resources will play a special role under the new condition of energy development. Concentration of practically the entire increase in fuel recovery in the eastern regions of the country and compensation of the decrease in operating capacity in the European basins at the expense of these resources will require a many-fold increase in the flow of energy resources, primarily natural gas, from east to west over distances up to 4000 kilometers. Realization of these flows will lead to the optimal development of all forms of transportation which will require creating (using the most advanced technological base) an overland energy transportation system unprecedented in capacity and extent. The problems of production allocation should also be solved in a new way, i.e., minimizing the distance between the energy sources and not only highly but also moderately energy-intensive enterprises.

Summing up the above said, the USSR energy complex is indeed entering the next transitional period, characterized by large-scale substitution of hydrocarbon fuel by more abundant energy resources: nuclear power and, in the future, coal. The related great restructuring of the energy balance in itself is not anything new for Soviet energy. Indeed, in the Soviet energy production, a quite rapid process of substituting coal for wood and low-grade local fuels occurred in the 1930–1950s, while in the 1960s–1970s, coal in its turn was vigorously replaced by oil and gas. As is evident from Fig. 10, the relative depth of structural changes in the energy balance at the first two stages of energy development in the USSR at least was not less than that planned for the future. But, above all, the forthcoming restructuring is characterized by much larger absolute sizes. Most important, in the past changes of structure, replacement of energy resources by new ones was accompanied by a lowering of their cost, while the forthcoming changes will involve not less than a doubling of the cost of primary production and, for this reason, cause special difficulties.

Under these conditions, the optimum energy strategy should not be universal in time. In the forthcoming transitional period, it is necessary to single out at least two phases: in the first phase, replacement of oil by forced recovery of natural gas should provide the time to develop large-scale nuclear power and increased coal extraction; then in the second phase, these energy resources will provide for further substitution of oil (even with a decrease in its production), and stabilization of the levels of natural-gas production as well.

3.2. The Role of Natural Gas in the Energy Balance of the USSR

Energy development in the USSR during the first phase of the transitional period is detemined primarily by the resolutions adopted at the 26th Congress of the Communist Party of the Soviet Union. As established by the plan for economic and social development of the national economy in the period 1981–1985 and by the basic directives up to 1990, it is directed to solving the following basic problems:

satisfying the demands of the national economy for fuel and energy, corresponding to an increase of 18% in the national income by 1985 and at least by a factor of 1.4 by 1990 compared to 1980;

elimination of the explicit and implicit deficits in the energy balance; creation of the required reserves of fuel and production capacity to secure normal reliability and quality of fuel and energy supply;

further increase in exports of energy resources compared to the high level in 1980 with a change in its structure in favor of increased exports of natural gas at the expense of oil and petroleum products.

The solution of these problems will require, first, great efforts in all sectors of the economy on conservation of energy and, second, large changes in the structure of primary energy production primarily in the sector (technologies).

The most important of these in the first phase will be the maximum substitution of natural gas for petroleum fuel.

The greatest possibilities for this occur in the substitution of gas for fuel oil in power plants. In 1980, approximately 125 million tons of oil were burned here, which constituted more than one-third of the total fuel consumption by power plants. Of this amount, more than one-third was burned by power plants which were completely adjusted for burning natural gas as well. Displacement of this fuel oil will not require any expenditures, but it does depend on the availability of natural gas and the gas storage capacity. The point is that mainly these power plants regulate the seasonal nonuniformity of gas consumption, using its surplus in summer and switching to oil in winter, when the gas consumption for heating sharply increases. According to calculations, in order to replace the indicated amount of oil in 1980, it would have been necessary to double the active gas storage capacity. In addition, during the whole studied period it is not efficient to replace fuel oil in power plants that cover the daily peak needs.

A considerable amount of oil could have been replaced by natural gas with relatively small expenditures on laying branches from the main gas pipelines and strengthening the gas economy. Taking this into account, with the availability of replaced energy resources, the oil consumption by power plants could have been decreased in 1980 by a factor of 1.7. In addition, part of the fuel oil could have been replaced by gas in boiler plants with the corresponding increase in gas storage capacity and with saving the oil as a fuel for the peak periods. For industrial oil consumers, the possibilities for replacing their oil without significant expenditures on reconstruction of the consumers presently constitute about 15% of total oil recovery.

The actual dimensions of such substitution are determined by the possiblities of increasing the throughput capacity of the main gas pipelines from the north of Western Siberia to the central regions of the country, as well as by gas storage capacities. The possibilities of increasing the secondary oil refining capacity, in which the heavy (oil) fractions can be refined into white petroleum products, are even greater. Without this, the replacement of oil by other forms of fuel makes no sense, since its use by the remaining consumers involves higher costs than by power plants.

Thus the immediate measure on changing the productive structure of the energy complex during the first phase is a fundamental reconstruction of the oil refining industry, directed toward considerable increase in the depth of petroleum refining. Solving the problems of substituting gas for oil, this must be done in the most economical manner. This is made possible by the territorial distribution of oil-fired power plants, in particular, their high concentration in Ural-Povolzh'e, through which large main gas pipelines pass from the north of Western Siberia. This opens up the possibility of using gas on the so-called short arm: continuation of the gas pipeline to Bashkiriya and Central Povolzh'e and replacing the oil at local power plants, using it first for transportation as an exchange for gas in the western regions of the country, and later for more extensive refinement at local petroleum refining plants. This manoeuvre made it clear that there was no need to cosntruct one of the large-diameter gas pipelines to the central regions of the country.

The replacement of oil at power stations is an enormous, but by no means the only measure involving the replacement of oil. Numerous measures on conservation of petroleum products must play an appreciable role here. The most important measure is an accelerated shift to diesel fuel of bus and truck transport. In addition, further electrification of railways, capable of stabilizing the consumption of diesel fuel with increasing turnover of freight on railways, is an important step. Another important step is the use of compressed natural gas in vehicles. At the same stage, refining the natural gas from gas-condensate fields in order to extract liquid fractions, which are a first-class raw material for petrochemistry, must be organized on a large scale.

The rapid increase in the volume of gas produced and transported requires intensified use of gas pipelines (in order to save pipes) by increasing the pressure of pumped gas and decreasing the distances between compressor stations, which rapidly increases the consumption of gas for the internal needs of gas pipelines. As a result, the gas industry is currently the third largest consumer of fuel (after power plants and ferrous metallurgy). If, on the other hand, the gas turbine drive for compressors is retained, then the consumption of gas for internal needs will increase twice as fast as total energy consumption.

Under the conditions of sharp increase in the costs of producing and transporting gas, the use of electric drive on gas pipelines becomes one of the most efficient measures in energy development. The most important candidates for this are the new main gas pipelines in regions with large interconnected power systems. Then, replacement of aging gas turbines and aircraft engines on the operating gas pipelines must play an increasing role. The calculations show that each ten billion kilowatt-hours used to drive compressors replaces 4–4.5 billion cubic meters of natural gas. A feature of the forthcoming stage is that nuclear power and heating cogeneration plants can produce additional electrical energy in the European part of the country. In this connection, less than 2 billion cubic meters of gas will be required to obtain 10 billion kilowatt-hours of electrical energy, while the remaining 2–2.5 billion cubic meters can be directed to other consumers. In addition, construction of cogeneration plants instead of boiler plants (in practice, primarily small boiler plants, requiring large capital investments and labor forces and having lower efficiency) will permit the saving of labor resources and up to 10% of the fuel consumed for heat production.

Estimation of the role of natural gas in energy development in the

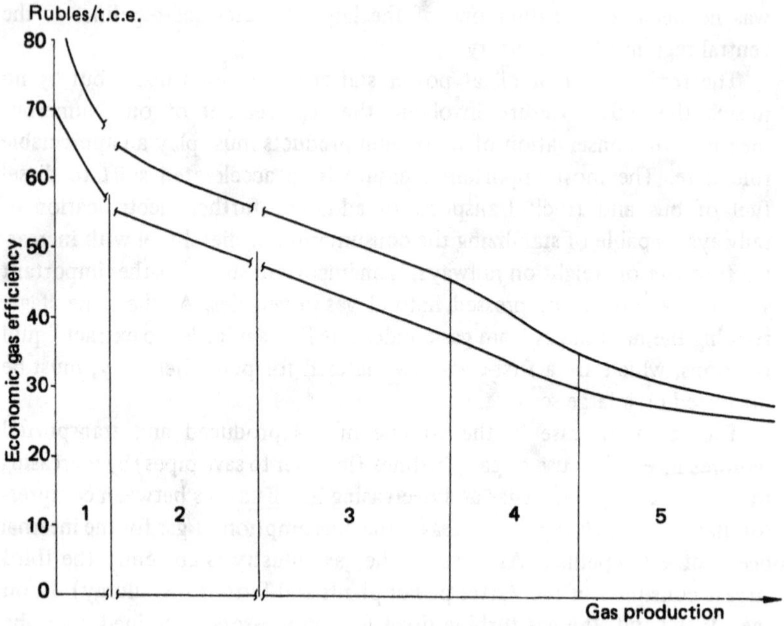

Figure 11 General form of the dependence of unit economic efficiency of gas on the amount of gas recovered. (1) Reserve measures for replacing gas. (2) Gas replacement with disassembly of operating power plants and in industrial installations. (3) Gas replacement by coal-fired cogeneration and nuclear boiler plants in district heating. (4) Gas replacement by coal and nuclear cogeneration plants in district heating. (5) Gas replacement by coal and nuclear energy in new power plants.

USSR during the second phase of the transitional period requires that the optimum dynamics of utilization of its limited resources during this period be determined first.

The optimization model of the dynamics of developing gas-bearing provinces was used for this investigation [7]. The main initial information for the model, as noted in Section 2.1, is a sequence (according to the final years of each five-year plan) of gas utilization efficiency functions that is obtained while studying the future energy development.

The general form of such a function is shown in Fig. 11. The least efficient area for gas utilization is for new power plants (condensing and cogeneration plants), where substitution of coal-fired or nuclear power plants (depending on region) for gas-fired plants costs, according to the levelized costs, from 25 to 35 rubles/t.c.e. If gas production decreases further, then it will be necessary to intensively replace the gas for purposes of supplying heat first by coal-fired boiler plants (in the regions with small load concentrations, as a rule, outside large cities where severe ecological restrictions on the use of coal arise) and nuclear cogeneration plants (NCP) (on the contrary, with the highest required load concentrations) where its use gives an economic effect ranging from 35 to 45 rubles t.c.e. Then, the new coal-fired cogeneration plants and nuclear boiler plants (NBPP) with equivalent costs of 50–55 rubles t.c.e. capable of competing with gas in a very wide range of average thermal load concentrations, become the gas-replacing installations. Further decrease in natural-gas production will require intensive disassembly of operating gas-fired power plants on supercritical steam parameters involving costs of 50 to 70 rubles t.c.e. The cost of replacing natural gas in most industrial furnaces and installations falls into the same range. The gas consumption for these purposes is so large that in the foreseeable future it can compensate any conceivable changes in gas production except, possibly, for the end of the second phase of the transitional period, when the necessity of rejecting gas as a substitute for liquid fuel and using synthetic liquid fuel from coal and shale, involving costs up to 100 rubles t.c.e. and higher, is not excluded.

The total cost function, presented in Fig. 11, is weighted against the gas production function which takes into account exploration, development and transportation costs. The gas production–cost function for increasingly commercially recoverable reserves is determined with the help of a stochastic model of the process of gas-bearing province [7] exploration; its general form is shown in Fig. 2. The cost of gas production is determined from a statistical analysis of technical-economic indicators of the development of operating and planned fields, which are first characterized according to their recovery conditions. Finally, expenditures on main transportation are determined from the projects of new gas pipelines, based on the location of each gas-bearing province, and the scheme for distributing its resources over the territory of the country, that is obtained by the energy complex optimization.

The results of multivariant study as the indicated complex of models of the optimum dynamics of gas production in one of the regions of the country were shown in Fig. 3. Attention should be paid to the good provision of initial gas production levels by the commercially recoverable reserves. They allow both retaining and increasing the starting production level by a factor of 1.7–1.8 and maintaining this level throughout the 20 year period. Taking into account the forecasted reserves, the starting production level can be increased by a factor of two (low alternative) or even by a factor of 2.5 (high alternative). But, in this case, the coefficient of confirmation of the forecasted gas reserves begins to play an important role. If this coefficient is taken to be close to unity, then it is possible to plan an increased production, since it can be maintained almost for 30 years from the starting year. But if the confirmation of recovery is low (due to inadequate geologic knowledge of the region, it can decrease to 0.5), the high production level will turn out to be unreliable, while approximately the same time intervals for maintaining the maximum level (approximately 30 years) can be secured only with decreased production.

It should be noted once again that the estimates of the maximum cost of gas production (calculated with the rent for depletion of the fields), obtained in the investigations, are insensitive to the production level achieved and remain approximately constant (increase very moderately) during the period of growth and stabilization of gas production, but begin to increase rapidly after the gas-bearing province enters the decreasing production phase. This means that stable and relatively moderate expenditures on gas in this region can be secured for a long period of time (in the example examined, up to 30 years from the starting year).

The over-all dynamics of gas production is formulated based on analogous investigations performed for the main gas-bearing provinces. The optimal production dynamics, corresponding approximately to identical reliability ranging from 0.85 to 0.95, is used here. In addition, new practically undeveloped and poorly explored gas-bearing provinces do not participate in the calculation of the over-all level of gas production during the first periods and they are viewed as sources for supporting the developed provinces as these provinces enter the decreasing production phase.

The prudent (from among the optimum) dynamics of gas production, formulated in this manner for the country as a whole, corresponds to the condition that the annual production levels remain constant during almost the entire second phase of the transitional period. In this case, it is possible to supply the most efficient gas consumers (with an economic effect from its use of up to 50–55 rubles t.c.e.) at a relatively moderate increase in the marginal costs of gas with the development of more expensive and remote gas-bearing provinces. Under favorable conditions (high coefficients of confirmation of forecasted reserves, accelerated technological gas production and transportation, etc.) it will be possible either to decrease the unit cost of gas or to increase production levels with the same costs, reserving the other, more inertial energy resources or singling out an additional amount of gas for export. It is this strategy that is contained in the efficient dynamics of changing the productive structure during the transitional period of energy development in the USSR, presented in Fig. 10.

3.3. Energy and Gas Supply for Consumers and for the Main Regions of the Country

The examined strategy of drastic shifts in the productive structure of the energy sector, permits continuing the qualitative improvement, begun in the 1950s, of the fuel and energy supply to the main categories of consumers. But, the principal means for this improvement will be nuclear power other than hydrocarbon fuel.

This is illustrated in Fig. 12. It is evident from this figure that the consumption of energy resources for nonfuel needs and as a raw material, as well as for small thermal units will be supplied as before only by fossil fuel, and increasingly by gas and synthetic liquid fuel. Fossil fuel will also retain its dominant role in industrial installations, but high-temperature nuclear reactors will begin to be used in the second phase of the transitional period: in ferrous and nonferrous metallurgy, chemical industry, etc. The increase in fossil-fuel consumption by end users will be almost completely secured by gas, while coal will remain here in the blast-furnace process (coke) and probably in the cement industry, at least in the eastern regions of the country.

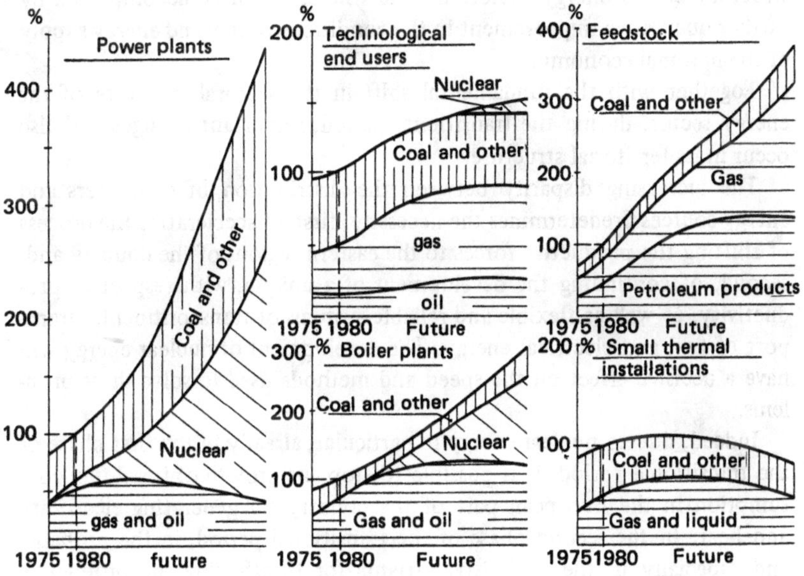

Figure 12 Structure of fuel supply for the main categories of consumers.

In contrast to this, the increase in demand for energy resources by boiler plants in the second phase of the transitional period should be met by nuclear energy (nuclear boiler plants and in the distant future possibly nuclear sources with dissociating heat carriers). As a result, in principle, it is possible to achieve stabilization, and then even some decrease in the consumption of gas and oil by boiler plants with very moderate growth in coal consumption (see Fig. 12). As a result, the share of high-grade resources (hydrocarbon fuel and nuclear energy) will increase in boiler plants from 70% in 1975 to approximately 90% in the second phase of the transitional period, which will completely solve the problem of the quality of energy supply to this category of consumers.

The only consumers for which coal will approximately retain its existing position in the future are power plants. Large-scale use of nuclear energy here will permit decreasing the consumption of gas and oil, but will not significantly alter the situation, when coal constitutes two-fifths of the consumed fuel. Thus, power plants will be the only consumers for which in the foreseeable future there will not be a fundamental improvement in the qualitative structure of the energy resources used. This is natural, since power plants are least sensitive to the quality of fuel and can adapt, with least damage, to the consequences of the required change in the productive structure of the energy sector. On the whole, it will be accompanied by further qualitative improvement in the conditions of fuel and energy supply of the national economy.

Together with the fundamental shift in the sectoral structure of the energy sector, during the transitional period, important changes will also occur in its territorial structure.

The increasing disparity between the distribution of consumers and energy sources predetermines the necessity, first, of accelerating the process of shifting the productive forces to the eastern regions of the country and, second, of continuing the development of a unique, with respect to productivity, as well as flexible and reliable systems of transcontinental transport of fuel and electrical energy. The development of nuclear energy will have a decisive effect on the speed and methods used to solve these problems.

Indeed, due to nuclear energy in particular, already in the first phase of the transitional period, it is possible to stop the growth of fossil-fuel consumption in the European part of the country for generating electricity and heat. In the second phase of the transitional period, in the probable and especially in the optimistic variants for transferring the productive forces to the eastern regions of the country, due to a vigorous energy conservation policy and the development of nuclear power, it is possible to solve in principle the key problem of improving the regional energy structure: to stop the growth, and then to begin a gradual decrease in the flows of all types of fossil fuel from the eastern regions of the country to the european regions. The solution of this problem would mean a very important landmark in the shift in the productive structure of the energy sector relative to changing conditions for its development, overcoming one of the main factors in increasing costs on energy development.

The other qualitative change in the regional transportation links of energy sector during the transitional period is the stabilization and subsequent decrease in the volumes of oil and, later, natural gas transported from eastern to european regions with continued growth in the transport of coal (the Kuznetsk and possibly processed Kansk-Achinsk coal) and rapid increase in flows of electrical energy (primarily from power plants on the Kansk-Achinski coal) and, in the future, of SLF. The formed scheme for regional links of energy complex in the second phase of the transitional period is shown roughly in Fig. 13.

The third feature of the regional structure of the energy sector in the forthcoming stage involves the change in the role of Central Asia and Kazakhstan in the energy balance of the country. The excess energy balance of this region supplying until recently one-third of the energy resources (primarily gas) to the European part of the country, will remain only in the first phase of the transitional period. In the forthcoming period, the expected growth in energy consumption in this region with gradual attainment of the maximum production levels for the basic energy resources will require, first, a serious solution of the problem of utilizing nuclear power here (which is complicated by the seismic activity of the region) and, second, finding the efficient schemes for replacing the flow of gas, delivered to the European regions along the existing pipelines by other energy resources. Such compensation can be secured by Siberian energy resources: the Tyumen gas in the eastern regions of Kazakhstan and coal for power plants in Central Asia. In addition, in the second phase of the transitional period, Central Asia and Kazakhstan will be transformed into a self-balanced energy region with equal volumes of exported and imported energy resources. The possibility that this region will become a net importer of energy resources in the future is not excluded and this must be taken into account when solving the problems of locating large energy-intensive industries here.

In the considered time horizon the energy balance of Siberia and the Far East will be most dynamic. The rapid growth of energy consumption

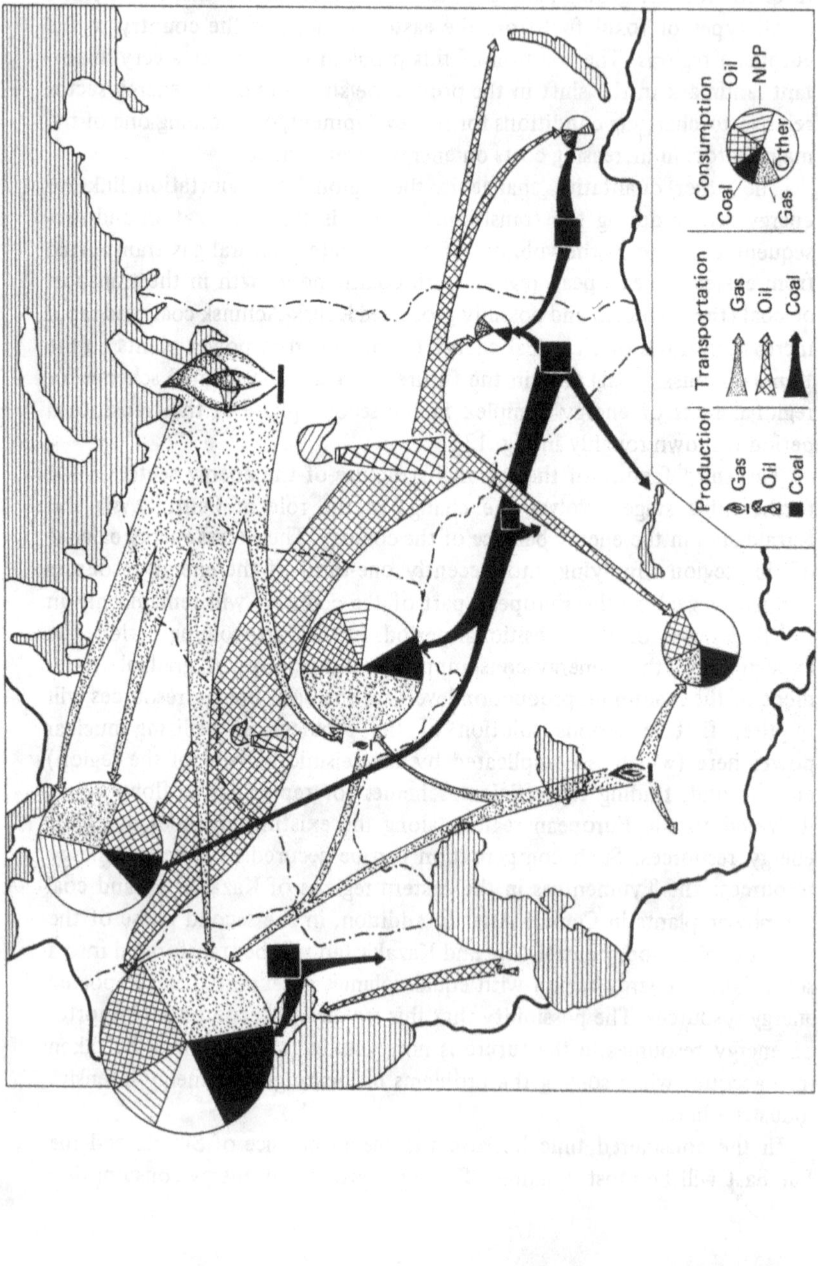

Figure 13 Diagram of fuel and energy flows.

in this region is accompanied by an even more rapid growth in primary energy production. Only under this condition it will be possible to provide the required export of energy resources out of Siberia to the European part of the country, and then to Central Asia and Kazakhstan. The structure of their production in this region is also changing rapidly. If in 1975 oil and gas production constituted approximately 55% of the overall volume of the local energy resources, in 1980 it already increased to 72% and later stabilized in the 75–80% range. This trend is reflected in the change in the strucutre of energy resources exported from Siberia. At the 1975 level, the export of oil and gas was 72% of the total volume of exports of energy resources and by 1980 it increased to 88%, which will remain approximately constant in the future.

The relation between different directions of transporting energy resources out of Siberia will be changing considerably in the future. In the considered time horizon, transportation of energy resources from Siberia to the European part of the country will remain the prevailing direction, but its share in the total volume of exports from Siberia will gradually decrease from 96% at the present time to 85–90%. The relative significance transporting energy resources out of Siberia to Central Asia and Kazakhstan increases correspondingly.

The unprecedented qualities of energy resources transported over distances of 3–4 thousand kilometers require a corresponding development of the transportation network. For transportation of gas from Siberia to the European part of the country, the first phase of the transitional period is most complicated; later the volumes of gas transported are either stabilized or may even decrease. However, the problem of choosing the form of gas transportation method (methanol, cooled gas, etc.) will remain urgent in the second phase as well, since by this time it will be necessary to replace the ageing networks of gas pipelines constructed before 1980. Then we can expect the quantities of coal transported out of Siberia to the European part of the country to increase. The most efficient combinations of different forms of coal transportation (special coal-carrying main lines, transmission of electrical energy, etc.) must still be chosen.

Thus the decisive role of further development of the gas industry in the fundamental restructuring of energy production in the USSR follows from the new energy strategy examined above.

References

1. Lalayants, A. M., A. A. Makarov, and L. A. Melentiev (1980). "The GOELRO

Plan and the Development of the Fuel–Energy Complex of the USSR", *60 Let Leninskogo Plana GOÉLRO (60 Years of Lenin's GOELRO Plan)*, Énergiya, Moscow, pp. 77–88.

2. Makarov, A. A., and A. G. Vigdorchik (1979). *Toplivno-ėnergeticheskii Complex (The Fuel–Energy Complex)*, Nauka, Moscow.
3. Ryps, G. S. (1978). *Ékonomicheskie Problemy Raspredeleniya Gaza (Economic Problems in Gas Distribution)*, Nedra, Leningrad.
4. Kurnosov, I. P., T. I. Klokova, I. P. Krutikova, and Yu. L. Khechinashvili (1977). "Use of Natural Gas in the National Economy", *Ékonomika Gazovoĭ Promyshlennosti*, No. 11, 29–34.
5. Melentiev, L. A. (1979). *Sistemny Issledovaniya v Énergetike (Systems Analysis in Energy)*, Nauka, Moscow.
6. Melentiev, L. A., and A. A. Makarov (1980). "Future Development of the Energy Complex," *Planovoe Khozyaistvo*, No. 4, 87–94.
7. Golovin, A. P., V. I. Kitaĭgorodskiĭ, and I. Ya. Faĭnshteĭn (1979). "Forecasting of Reserves and Optimization of Regional Levels of Gas Production within the Energy Complex," *Ékonomika i Matematicheskie Metody* **15**, No. 5, 940–950.
8. "Methodological Principles of Economic Evaluation of Mineral Deposits," *Ékonomika i Matematicheskie Metody* **14** No. 3, 405–419 (1978).
9. Kitaĭgorodskiĭ, V. I., and I. Ya. Faĭnshteĭn (1981). "Problems of the Reliability of Raw Material Provision for Natural Gas Production," *Izv. Akad. Nauk SSSR, Énergetika i Transport*, No. 4, 20–31.
10. Makarov, A. A., "Systems Analysis of the Future Structure of the Energy Complex of the USSR," *Izv. Akad. Nauk SSSR, Énergetika i Transport*. No. 3, 21–30.
11. Tkachenko, G. E. (1979). "Forecasting the Conditions for Long-Range Development of Energy Complex Taking into Account the Uncertainties in its External Links," *Metody Analiza i Modeli Struktury Territorial 'no-proizvodstvennykh Kompleksov (Methods of Analysis and Models of the Structure of Regional Production Complexes)*, Nauka, Novosibirsk, pp. 47–59.
12. *Narodnoe Khozyaistvo SSSR v 1980 Godu (The National Economy in 1980)*, Statistika, Moscow, (1981).
13. *Metododicheskie Polozheniya Optimizatsii Razvitiya Toplivno–énergeticheskogo Kompleksa (Methodological Aspects of Optimizing the Energy Complex Development)*, Nauka, Moscow, (1975).
14. Kononov, Yu. D., A. G. Korneev, and V. Z. Tkachenko (1979). "Simulation of the External Production Links of the Industrial System", *Ékonomika i matematicheskie metody* **15**, No. 5, 969–977.
15. Kononov, Yu. D. (1981). "External Production Links and the Inertness of Energy Complex Development", *Izv. SO Akad. Nauk SSSR, Ser. obshchestvennykh nauk* **6** No. 2, 12–18.
16. Dobrovol'skiĭ, G. P., S. M. Klimenko, Yu. A. Kuznetsov, and V. I. Rabchuk (1973). "Analysis of the Economic Efficiency of low-temperature Gas Pipelines", *Proektirovanie i stroitel'stvo truboprovodov i gazoneftepromyslovykh sooruzheniĭ*, No. 5, 34–40.
17. Furman, I. Ya. (1978). *Ékonomika magistral'nogo transporta gaza (Economics of Main-Line Gas Transportation)*, Nedra, Moscow.
18. Makarov, A. A., and L. A. Melentiev (1981). "Problems and Paths in Energy Production in the USSR", *Ékonomika i organizatsiya promyshlennogo proizdvodstva* No. 3, 17–45.

Sov. Tech. Rev. A Energy Reviews, Vol. 2, 1985, pp. 53–109
0275–7893/85/002–053 $30.00/0

BASIC TRENDS IN SCIENTIFIC AND TECHNICAL PROGRESS IN MAINLINE GAS TRANSPORTATION

A. D. SEDYKH and Z. T. GALIULLIN

Ministry of the Gas Industry of the USSR, 117939, Moscow, B-31, Ulitsa Stroitelei̇̆, 8, Korpus 1

All-Union Scientific-Research Institute of Natural Gases, 142700, Moscow Province, Vidnoe, GSP

Abstract

Scientific and technical progress in mainline gas transportation in the USSR is examined and its effectiveness is analyzed. The emphasis is on solutions to technical problems that help to realize the policy for supplying energy.

The methods currently used to calculate the reliability of gas transportation systems are formulated and their use in designing the Ureng–Pomar–Uzhgorod gas pipeline is discussed as an example.

Contents

C

1. Basic Characteristics of the Development of Pipeline Gas Transportation

The gas industry in the USSR is one of the key industries in the energy complex, determining to a large extent the technical progress and rate of development of the entire economy. In 1980, gas production exceeded 435 billion cubic meters. The relative contribution of gas to the energy balance of the country is rapidly increasing: 27% in 1980 as opposed to 7.9% in 1960 (see Table I).

The distribution of the raw material supply and of the principal gas consuming regions of the USSR made it necessary to construct a branching network of main gas pipelines, capable of providing an uninterrupted flow of gas to the national economy.

The development of the gas transportation subsector is distinguished by its size and dynamism. More than 136 thousand kilometers of main gas pipelines, 314 compressor stations, and more than 2100 gas distribution stations are currently in operation in the gas transportation systems. Practically the entire volume of gas produced in this country is pumped along these gas pipelines.

Main gas pipelines with a total extent exceeding 30 thousand kilometers and compressor stations with a total capacity of 8.6 million kilowatts were constructed and put into operation only during the 10th Five-Year Plan, which provided for an increase of 145.9 billion cubic meters in gas production during these years.

During the last 20 years, in spite of a 3.14-fold increase in the average transportation distance, the volumes of gas transported increased more than 15-fold, while the length of main gas pipelines has increased only 6.27-fold (see Table I).

Table I Basic Indicators of Mainline Gas Transportation (at the End of the Year)

Indicators	Units	Years				
		1960	1965	1970	1975	1980
Gas production	billion cubic meters/year	45.3	127.7	197.9	289.3	435.2
Relative contribution of gas to the energy balance of the country	%	7.9	15.5	19.1	21.8	27
Gas (commercial) flow to the economy	billion cubic meters/year	26.0	112.1	181.5	279.4	401.1
Length of main gas pipelines	thousand kilometers	21.0	42.3	67.5	98.7	131.6
Installed capacity of gas pumping aggregates	thousand kW	270	2100	3900	9000	17,600
Average distance of gas transport	km	589	656	917	1237	1851
Relative fraction of gas pipelines with diameter						
1020, 1220 and 1420 mm	%	3.2	17.9	21.2	39.9	46.6
1220 and 1420 mm	%	–	–	–	19.3	27.7
1420 mm	%	–	–	–	3.7	10.9

During the period examined, significant and qualitative changes have occurred in the structure of the natural-gas network. At the beginning of the 1960s, separate gas pipelines supplying gas to separate regions of the country were constructed. The characteristic feature of gas-pipeline construction in subsequent years is the creation of a long-range gas-supply system, which, as the gas and gas-condensate fields, gas refineries, and underground gas storage facilities came on line, was combined into a unified gas supply system of the USSR (UGSS). The creation of the UGSS formed the foundation for maximum utilization of the production capacities of gas production, gas transportation, and gas refining enterprises. Their interaction and mutual backup created flexibility of maneuvering gas flows and increased the reliability of the gas flow to consumers.

The high values of the indicators of the development of the gas-transportation subsector were achieved by realizing a long-term program of extensive assimilation of scientific and technological progress. Scientific-technical progress in subsequent years occurred for the following reasons:

increase the diameters of main gas pipelines;

increase in working pressure;

cooling of transported gas;

increase in the unit power, efficiency, and reliability of gas pumping aggregates (GPA);

increase in the reliability of main gas pipelines as a whole, etc.

The increase in the diameters of main gas pipelines was one of the basic paths for improving all technical-economic indicators of long-range gas transportation. In addition, the capacity of gas pipelines increased and the specific input of metal and capital (per unit volume of gas transported) decreased. All of this ultimately decreased the specific labor costs and reduced mainline construction and operating costs. This path of technical progress was successfully followed throughout the entire history of the development of the gas industry.

Thus main gas pipelines with diameters of 1020, 1220, and 1420 mm were constructed and operated in the USSR for the first time in the world. At the present time, pipes with the indicated diameters have been utilized in about 50% of all pipelines constructed. In spite of the comparatively short length of 1420-mm gas pipelines (approximately 11% of the total length), the volume of gas transported by such pipelines constitutes almost 50% of the total volume of gas transportated.

The transition at the end of the 1960s to the construction of main gas pipelines using 1220-mm pipe helped to increase their capacity by a factor of 1.4–1.5 and to decrease the specific input of metal by 8–10%. The increase in pipeline capacity accompanying an increase in diameter from 1020 to 1420 mm (excluding the effect of increased pressure and cooling of the gas) is estimated to be approximately a factor of 2.4; the decrease in the specific input of metal is approximately 15–18%.

The working pressure was increased in the USSR from 55 to 75 kg cm^{-2} in 1972. This additionally increased gas-pipeline capacity (in addition to the increase due to the use of pipes with larger diameters) by 33–35% and decreased the specific capital investment and reduced costs (due to lower specific volumes of construction-assembly work, use of better steels for the pipes, lower energy input per unit of gas transported) by approximately 5–6%.

As the diameters and working pressures increased, the influence of temperature factors increased to such an extent that it is practically impossible to transport gas in 1220- and 1420-mm gas pipelines without cooling it. For this reason, at the end of the 1960s, special investigations were made into the establishment of optimal cooling levels in gas pipelines in different environmental and climatic regions of the country. As a result, it is established that depending on the environmental and climatic conditions for laying and operating gas pipelines, it is useful to separate the cooling of transported gas into two levels:

cooling of gas to temperatures close to the ambient air temperature; it was demonstrated that the maximum effect is achieved when the gas is cooled to temperatures exceeding the ambient air temperature by 8–12°C [1];

cooling of gas to temperatures close to the temperature of the soil at the depth at which the gas pipelines are laid [2].

The first level was recommended for and realized on gas pipelines with diameters exceeding 1020 mm under all environmental and climatic conditions. The annual productivity increased by 6–8% and the specific reduced costs decreased by 2–3%.

Construction of gas pipelines in Western Siberia, which is characterized by the presence of unstable permafrost soils, makes it necessary to solve complicated technical problems, one of which is the elimination of thermal interaction between the gas pipeline and the surrounding medium. Investigations have show that an effective method here is cooling of the transported gas to the soil temperature (to minus 1–4°C). An efficient refrigeration cycle has been developed for this purpose using a mixed cooling agent (propane–butane mixture), which greatly (1.5–2-fold) reduces the heat exchange surface of the condensers, the most metal intensive parts of the refrigeration setup. Installations have now been developed based on the indicated cycle with a refrigeration capacity of 8 and 16 million kcal per hour at a temperature level of -10°C driven with an electrical engine and with an aircraft-type gas turbine [3].

On the whole, relative to 1020-mm pipelines pressured at 55 kh cm^{-2}, the use of 1420-mm pipes at a working pressure of 75 kg m^{-2} with the transported gas cooled to the optimal level has increased pipeline capacity from 10 to 36 billion cubic meters per year, i.e., 3.6-fold, decreased the specific (per unit volume of gas transported) input of metal by 25%, and decreased the specific capital investment by 35%.

With respect to the stages of introduction of scientific-technical progress into long-range gas transportation, the evolution of the gas industry as a whole can be divided into three periods.

The first period (1960–1968) was the period of construction of main gas pipelines using 1020-mm pipes. Here, due to the two-fold increase in the pipeline capacity, related to the transition to 1020-mm pipes (replacing 720- and 820-mm pipes) in the period from 1960 to 1965, high rates of growth were achieved in gas production. Thus the increment to gas production increased from 9.9 billion cubic meters in 1960 to 19.1 billion cubic meters in 1965, i.e., almost two-fold. Then, in the period from 1966 to 1968, in spite of the increase in the rates and the scales of introduction of new pipeline capacity (from 4.26 thousand kilometers per year in the five year period from 1960 to 1965 up to 5.04 thousand kilometers per year in the period 1966–1970), the rates of growth of gas production in the country decreased from 19.1 billion cubic meters in 1965 to 11.7 billion cubic meters in 1968. During these years, the increase in the range of gas transportation was not compensated by an increase in the rates of pipeline construction and the achieved level of technical progress. For this reason, preparations were made for the next stage of introduction of scientific and technical progress (see Fig. 1).

The second period (1968–1972) was the period of construction of main gas pipelines with 1220-mm pipe. At the beginning of this period (1969–1970), the 1.4–1.5-fold increase in the capacity of gas pipelines was still compensated by the increase in the transportation distance and provided for an increase in the rate of growth of gas production from 11.7 billion cubic meters in 1968 to 16.8 billion cubic meters in 1970. However, the average annual increase in gas production nevertheless turned out to be lower than the increase in production in the period 1960–1965. And, in the following years 1971–1972, even a 1.24-fold increase in the volume of pipeline construction did not permit maintaining these increments to gas production. The increase in gas production

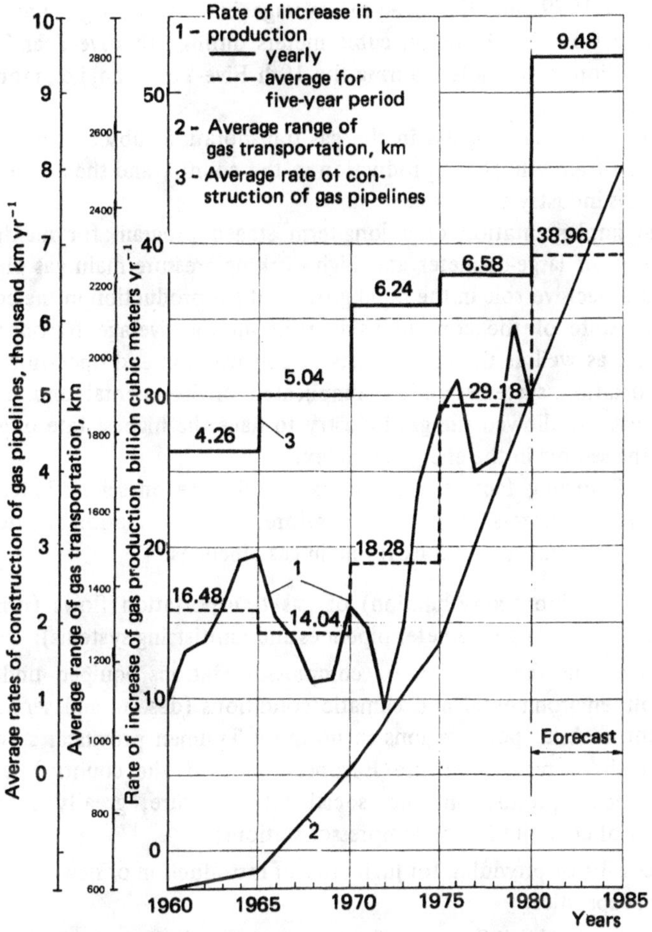

Figure 1. Dynamics of basic indicators of long-range gas transportation in the USSR.

decreased from 16.8 billion cubic meters in 1970 to 9.0 billion cubic meters in 1972. A new, greater increase in the capacity of gas pipelines was unavoidable during this period.

The third period (1973 to present) is the period of construction of main gas pipelines with 1420-mm pipe pressurized at 75 kg cm^{-2} and with cooling of the gas.

This stage of technical progress gave a new strong impetus toward a sharp growth in gas production. The annual increment to gas production increased from 9.0 billion cubic meters in 1972 to 34.4 billion cubic meters in 1979 and the five-year average increment to gas production increased from 18.28 billion cubic meters during 9th Five-Year Plan to 29.18 billion cubic meters during the 10th Five-Year Plan, i.e., more than 1.6-fold.

Thus technical progress in the gas transporation subsector determines the rate of growth of gas production in the country and the development of the gas industry as a whole.

The implementation of a long-term staged program for making the transition of large-diameter and high-working-pressure main gas pipelines played a decisive role in the rapid growth of gas production in the country and, in spite of the continuous increase in the average transportation distance, as well as the higher costs of construction and operation of gas transportation systems under complicated environmental and climatic conditions, it allowed the gas industry to have the highest rate of growth of all the sectors in the energy complex.

The following factors had a decisive influence on scientific-technical progress in compressing gas and, therefore, on the formation and development of the compressor stock for main gas pipelines:

concentration (consolidation) of gas transportation flows (primarily construction of large-diameter pipelines and multistring systems);

increase in the fraction of compressor stations equiped under the difficult environmental and climatic conditions (deserts and semideserts in Central Asia, polar regions in northern Tyumen province, as well as the northern regions of the European part of the country) with an undeveloped production and social infrastructure, greatly increasing the cost of construction of compressor stations;

necessity of providing for high rates of introduction of new compressor station capacities;

increase in the influence of equipment reliability on the indicators of long gas pipelines;

increase in fuel-energy consumption for the internal needs of compressor stations;

increase in the requirements for booster compressor stations.

The concentration of gas flows in pipelines led to a continuous increase of not only the maximum unit capacity of GPA, but also of the average unit capacity of the entire stock of gas pumping aggregates. During the last ten years (1970–1980), the average unit capacity of GPA increased from 5 to 7.83 thousand kilowatts, while the maximum capacity increased from 10 to 25 thousand kilowatts (see Fig. 2 and Table II).

Table II Dynamics of Variation of Indicators of the Gas Pipeline Stock (at the End of the Year)

Indicators	Year		
	1970	1975	1980
Average unit capacity, thousand kW	5.0	6.3	7.83
Maximum unit capacity, thousand kW	10.0	16.0	25.0
Mean-weighted (according to capacity) efficiency of stock, %	24.8	26.2	26.4

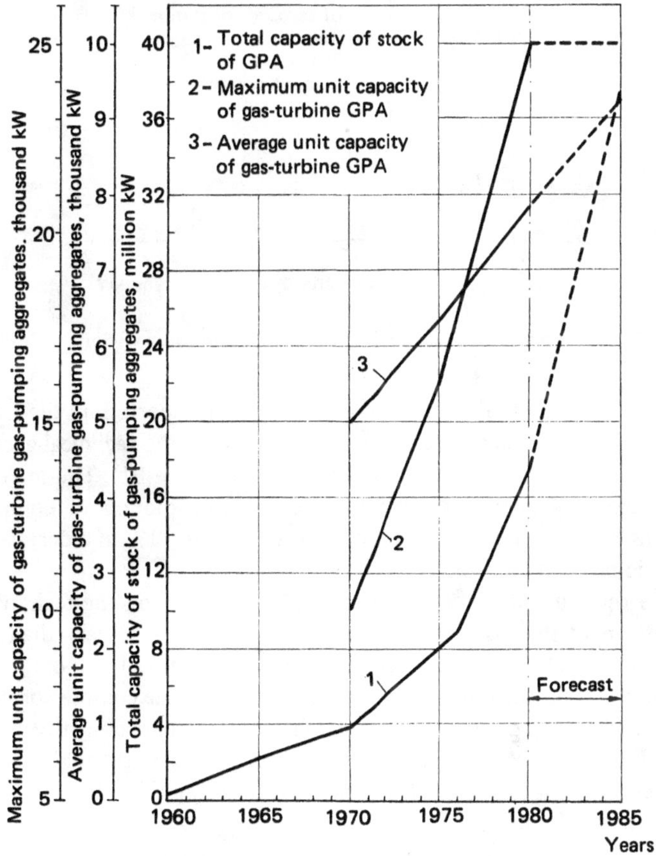

Figure 2. Dynamics of stock of compressor stations power drive.

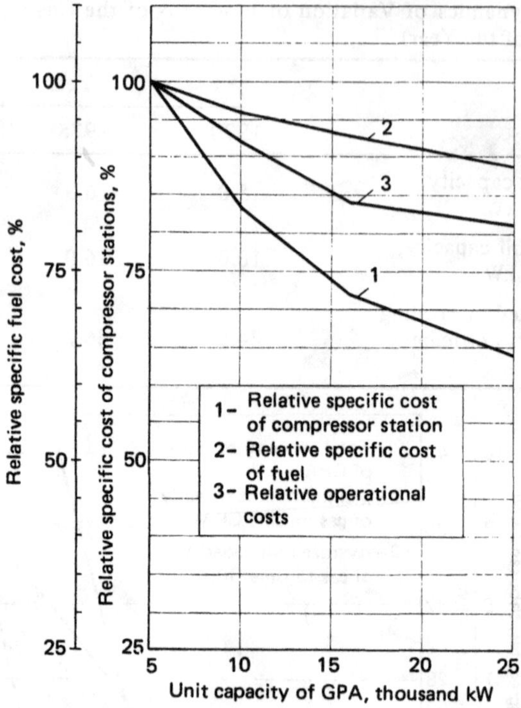

Figure 3. Relative change in the indicators of compressor stations with identical level of reserves.

As in other industrial sectors, in pipeline gas transportation, the increase in the unit GPA capacities led to an increase in the technical-economic indicators of compressor stations (Fig. 3 and Table III). The increase in unit capacities decreases the unit cost of construction of compressor stations by 36%, decreases the unit cost of fuel by 11%, and decreases the operational costs by 19%.

The optimum level of increase of GPA depends on the technological parameters of the gas pipeline (compressor station capacities), the reliability of GPA, and the length of the gas pipelines. Technical-economic studies have shown that the optimum standard size of pipeline compressor stations for a single-string gas pipeline can be obtained by connecting three operating GPA with the appropriate capacity in parallel with full-head superchargers. On the other hand, the number of backup plants is determined by the reliability of the gas supply.

Thus, for example, optimum compressor stations on 1420-mm mainlines pressurized at 75 and 100 $\mathrm{kg\,cm^{-2}}$ are based on gas pumping aggregates

Table III Relative Change in the Specific Cost of Fuel and Operational Expenses for Compressor Stations with an Identical Level of Reserves

Indicators	Unit capacity of gas-pumping aggregate, thousand kW			
	6	10	16	25
Relative unit cost of compressor station, %	100	84	72	64
Relative unit cost of fuel %	100	96	93	89
Relative operational costs, %	100	92	84	81

with unit capacity of 16 and 25 thousand kilowatts, respectively, with three working and one backup plant. In this manner, it has been established that for gas pipelines with diameters of 1020, 1220, and 1420 mm pressurized at 56, 76, 100, 120 kg cm^{-2}, the optimum compressor station structures can be obtained from a standard series of gas pumping aggregates with capacities of 4–6–10–16–25 thousand kilowatts. The efficiency of these plants is also confirmed in terms of the GPA capacity.

The technical-economic indicators of the efficiency of compressor stations with gas-turbine and electrical drive (including the cost of electricity) are close to one another. The data in Table IV and Fig. 4 illustrate the change in the relative fraction of different types of GPA drives from 1960 to 1980.

High rates of introduction of new capacity can be provided for only by significant simplification and industrialization of the construction-assembly of compressor stations. The problem was solved by developing and introducing into practice nonregenerative, no-basement gas turbine installations (GTI) of stationary, marine, and aircraft types; increased use of block design and factory preparation of the main and auxiliary equipment; and, extensive use of the full-equipped–block method of construction based on the use of standardized, easily erected structures. Gas turbine plants with aircraft and marine engines, which constituted 16.7% of the overall stock of GPA by the end of the 10th Five-Year Plan, best meet the requirements of industrialization of the construction of compressor stations. At the same time, acceleration (by a factor of 1.5–2) and lower cost of construction of compressor stations using nonregenerative GTI (by 10–12%) stopped, for a certain period (before the end of the 11th Five-Year Plan), the progressive increasing trend in the economic utility

Table IV Dynamics of Change of the Compressor Stock and its Structure (at the End of the Year)

Indicators	Years				
	1960	1965	1970	1975	1980
Total capacity of gas-pumping aggregates, million kW	0.27	2.1	3.9	9.0	17.6
Relative fraction of drive, %					
gas-turbine	33	35	56	70	77.3
electrical	37	53	30	19.4	17.0
piston	30	12	13	9.6	5.7

of the stock of gas turbine plants with respect to the unit consumption of fuel at an effective efficiency of 26–27%. The technical progress in the technology of compressing gas is scientifically well founded [4, 5].

This technology, provides for the following:

application of full-head centrifugal superchargers (CFS) and transition to parallel scheme for plant operation at compressor stations;

modification of CFS with a replaceable flow-through part (RFTP in order to increase the efficiency of pipeline compressor stations and the operational efficiency of installed capacity.

The program of accelerated development of gas production, primarily, due to accelerated assimilation of gas fields in Western Siberia, is assured by the basic directions of economic and social development of the USSR over the period of 1981–1985 and up to 1990, which were consolidated by the 26th Congress of the Communist Party of the Soviet Union. By 1985, gas production should reach 600–640 billion cubic meters. Higher production of Siberian gas and the solution of the problem of transporting it to the European part of the country are the most important links in the energy program of the 11th and 12th Five-Year Plans.

An objective factor, which makes possible the realization of the planned program, is the large supply of raw material, developed in our country [6].

At the same time, it should be noted that the structure of gas resources has a number of important characteristics which determine a number of problems in the development of the industry. First of all, we should point out the very specific geographical distribution of gas resources. Of the overall commercial reserves, only somewhat more that 12% occur in the European part of the country, while 88% occur in the Asian regions.

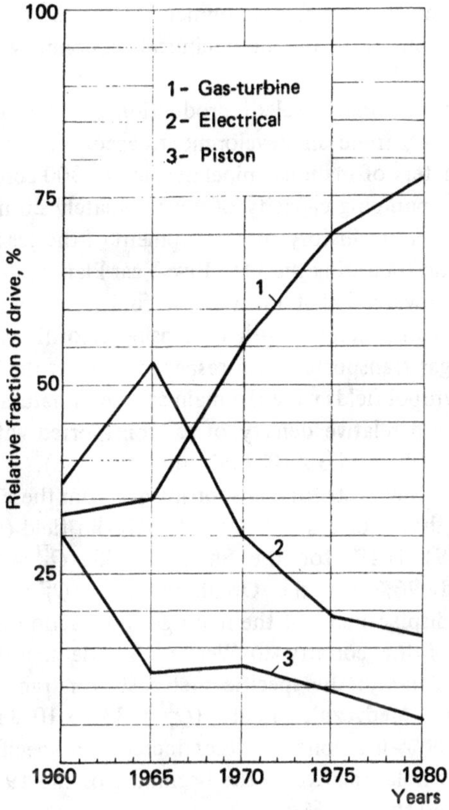

Figure 4. Structure of compressor stock in the USSR.

Most of the explored reserves of gas (74%) are confined to Western Siberia, primarily to its northern regions, access to which is difficult. Such a geographic distribution of gas resources cannot be viewed as being favorable, since most of the resources are far removed from regions where the gas is utilized.

The displacement of the raw material supply of the industry to the northern regions of Tyumen province makes it necessary to construct main gas pipelines under difficult environmental and climatic conditions and sharply increase the distance over which gas is transported. In 1980, the average transportation distance was 1850 km, but by 1985, according to forecasts, it will exceed 2500 km, while the length of separate gas pipelines will reach 4000–4700 km. The construction of gas pipelines in the north and the increase in the gas transportation distance are objective factors of present and future development of the industry, which have a negative effect on most of the technical-economic gas-transportation indicators.

To ensure the planned level of production and transportation of gas with the emerging trend in development, according to forecasts, 47.4 thousand kilometers of main gas pipelines, about 300 compressor stations with a total gas pumping capacity of approximately 20 million kilowatts – exceeding the entire history of development of the gas industry – must be put into operation during the 11th Five-Year Plan.

It is well known [7] that the specific fuel costs, per unit of freight transportation work, increase with decreasing calorific value and relative density of the gas transported with respect to air. At the beginning of the 1960s, the Stavropol fields were the main source of natural gas (the lowest calorific value and relative density of gas transported with respect to air are, respectively, $Q_P^H = 35.6 \times 10^6\ \mathrm{J\,m^{-3}}$ and $\Delta = 0.56$).

If the specific fuel costs for transporting gas from these fields are scaled as 100%, then these costs are 98% for the Gazlin field ($Q_P^H = 34.4 \times 10^6\ \mathrm{J\,m^{-3}}$, $\Delta = 0.59$) 101% for the Shatlyk field ($Q_P^H = 33.1 \times 10^6\ \mathrm{J\,m^{-3}}$, $\Delta = 0.56$), and 96% for the Orenburg field ($Q_P^H = 33.9 \times 10^6\ \mathrm{J\,m^{-3}}$, $\Delta = 0.6$). The displacement of the main gas production regions from the southern part of the country to Western Siberia increases the specific fuel costs considerably: the specific fuel costs for transporting gas from the Ureng and Medvezh'e fields ($Q_P^H = 33.1 \times 10^6\ \mathrm{J\,m^{-3}}$, $\Delta = 0.56$) increased by 109%; the concommitant increase in specific fuel costs for the UGSS as a whole, relative to the beginning of the 1970s, is estimated (at the 1985 level) to be 5–6%.

All this leads to a further increase in the consumption of energy for transporting gas over main pipelines. At the present time, the consumption of gas for the internal needs of gas pipelines is of the order of 31 billion cubic meters per year. If the economic efficiency of main gas pipelines is not increased, then by the end of the period under study, gas consumption for internal needs could reach 70–75 billion cubic meters per year.

An analysis of the effect of a change in the intial cost indicators on the optimum parameters of gas pipelines based on the traditional gas-transportation technology utilizing a 1420-mm pipeline pressurized at 75 kg cm^{-2} showed the following.

The optimum capacity of gas pipelines decreases with the increasing cost of fuel gas and installed capacity of compressor stations and increases with increasing cost of laying gas pipelines. The optimum capacity of gas pipelines is affected most by a change in the cost of 1 km of gas pipeline and 1 kW of compressor station capacity. An increase in the cost of fuel gas does not affect the optimum capacity as strongly. Nevertheless, the rising cost of fuel gas could become a determining factor for the following reasons. The opposing influence of the first two factors leads to a certain mutual compensation in the change of the optimal capacity. But, the expected range of the relative change in the cost of fuel gas greatly exceeds the possible change in the cost of constructing the linear part and the compressor stations of main gas pipelines. The possible relative change in the cost of 1 km of gas pipeline and 1 kW of compressor station capacity lies in the range 0–100%, while the relative change in the cost of fuel gas can be 1000% and more. For a relative change in the starting cost indicators by an amount exceeding 500%, the optimum capacity will be determined only by the price of fuel gas.

Therefore, the optimum capacity of gas pipelines should decrease in the future due to a possible increase in the cost of fuel gas. For example, if the price of fuel gas increases from 8–10 rubles/1000 cubic meters (present price) to 50–60 rubles/1000 cubic meters (double the price of the marginal† fuel), then the optimum productivity would decrease from 30–32 billion cubic meters per year to 24–26 billion cubic meters per year, i.e., approximately by 20%.‡ This will increase the stock and material required for new pipeline construction. In this connection, the significance of the restrictions on the material and labor resources required for constructing and operating main gas pipelines, as well as the reliability of their operation, will increase.

Thus the traditional technology (75 kg cm^{-2}, 1420 mm) will not allow for the increased gas flows required in the future. The technologies in the future must be based on the search for solutions to technical problems that increase the optimum unit flow of gas by not more than 20–25% with an acceptable level of consumption of energy and metal.

The objective characteristics of the development of gas pipeline transportation listed above predetermine two basic directions for scientific-technical progress:

intensification (consolidation) of unit gas flows;

lower consumption of energy and material by gas pipelines, and lower labor costs for pipeline construction and operation.

---

†Translator's note: the term "marginal," as used here, is defined in

‡In addition, if the degree of compression remains unchanged, then the spacing between compressor stations will increase from 110–120 km to 160–190 km or if the spacing between compressor stations remains unchanged, then the degree of compression must decrease from 1.52 to 1.42. The latter is the most economical choice.

These directions impose contradictory requirements on the optimum gas pipeline parameters. The choice of optimal gas transportation schemes must be made on the basis of a compromise, taking into account the change in the starting cost indicators. In addition, depending on the technical-economic state of affairs, at different stages of development of gas transportation, preference could be given either to intensifying the flows of gas or decreasing the amounts of energy and material consumed by gas pipelines.

For this reason and All-Union Scientific-Research Natural Gas Institute (VNIIGAZ) and other scientific-research and design institutes for the gas industry have analyzed all possible directions of development of pipeline transportation, including nontraditional directions [1, 8–11].

2. Technological Processes and Solutions to Technical Problems for Increasing Gas Pipeline Capacity

2.1. *Functional Model of Main Gas Pipelines*

The functioning of main gas pipelines can be described by the following system of equations:

$$q = a\,\frac{d_{\text{in}}^{2.6} P_i}{\sqrt{Z_{\text{av}}}}\sqrt{1-\frac{1}{\epsilon^2}}\ \text{million m}^3\,\text{day}^{-1}, \tag{1}$$

$$N = bqZ_{\text{intake}}\,T_{\text{intake}}\left(\epsilon^{(k-1)/r^k}-1\right)\text{MW}, \tag{2}$$

$$\delta = \frac{nPD_{\text{out}}}{2(R_1+nP)} = \frac{nPD_{\text{out}}}{2[(R_1^{\text{out}}m/K_1K_{\text{out}})+nP]}\ \text{mm}, \tag{3}$$

where

$$c = \frac{1.64\times10^6 E}{\sqrt{\Delta T_{\text{av}}L}}, \quad \epsilon = \frac{P_i}{P_f}, \quad b = \frac{0.00401K}{K-1};$$

N, and δ are the pipeline capacity, the working capacity of the compressor station, and the thickness of the pipe walls, respectively; P_i and P_f are the initial and final pressure on the section between compressor stations, respectively; d_{in} and d_{out} are the inner and outer pipe diameters, respectively; E is the coefficient of hydraulic efficiency of the gas pipeline; Z_{av} and Z_{intake} are the average coefficient of compressibility of the transported gas to the section between two compressor stations and the coefficient of compressibility at the compressor station intake, respectively; T_{av} and T_{intake} are the average absolute temperature of the transported gas for the section between two compressor stations and the temperature of the gas at the intake of compressor stations, respectively; and, L is the distance between two compressor stations.

For purposes of comparability, we shall assume that the relative specific weight of the gas with respect to air, the gas temperature, the distance between compressor stations, and the degree of compression, in terms of which the final pressure on the pipeline section is determined, are the same for all diameters and working pressure levels, keeping in mind that they are optimal values.

Then, according to Eq. (1), the throughput capacity of the gas pipeline is affected by the following parameters: the absolute initial pressure in the gas pipeline P_i, the inner diameter of the gas pipeline d_{in}, and the coefficient of compressibility of the gas z_{av}.

We shall use the following values of the coefficients entering into Eq. (3):

P, kg cm^{-2}	55	75	100	120	
	for 1420-mm pipe				
K_i	1.05	1.1	1.15	1.2	(3)
	for 1620-mm pipe				
K_i	1.1	1.15	1.20	1.25	

$n = 1.1$ is the coefficient of overloading of the working pressure in the gas pipeline: $R = 6000\,\text{kg cm}^{-2}$ is the working resistance of the pipe metal; $m = 0.9$ is the operating factor of the gas pipeline; $K_1 = 1.34$ is the safety factor for the material used.

Then the ratio of the capacity of the main gas pipelines with different diameters and initial pressures

$$\frac{q_1}{q_2} = \frac{P_{i1}\, d_{in1}^{2.6}}{P_{i2}\, d_{in2}^{2.6}} \sqrt{\frac{Z_{av1}}{Z_{av2}}}. \tag{4}$$

2.2. *Traditional Methods for Increasing the Capacity of Main Gas Pipelines*

2.2.1. Increase in Diameters of Main Gas Pipelines When the pipeline diameter increases from d_{in1} to d_{in2} (with other parameters remaining constant), the pipeline capacity increases from q_1 to q_2, according to Eq. (4), with

$$d_{in1} = D_{i1} - 2\delta_1; \quad d_{in2} = D_{i2} - 2\delta_2 \; (D_{i1} = 1420\,\text{mm}, D_{i2} = 1620\,\text{mm})$$

$$q_2 = q_2\left(\frac{d_{\text{in}\,2}}{d_{\text{in}\,1}}\right)^{2.6} = q_1\left(\frac{D_{i2} - 2\delta_2}{D_{i1} - 2\delta_1}\right)^{2.6} \approx 1.4q_1.$$

The specific consumption of metal (referred to unit pipeline capacity) for pipeline construction decreases from $M_{\text{sp}\,1}$ to $M_{\text{sp}\,2}$:

$$M_{\text{sp}\,2} \approx M_{\text{sp}\,1}\left(\frac{D_{i1}}{D_{i2}}\right)^{0.6} \approx 0.92M_{\text{sp}\,1}.$$

Thus, when the diameter increases from 1420 to 1620 mm, pipeline capacity increases by ~35–40% and the specific metal consumption decreases by ~7–8%. The specific fuel costs are independent of diameter for the restrictions adopted [7]. At the same time, pipeline reliability in flooded soils decreases due the increased buoyancy of the pipe (pipe buoyancy increases by ~30%). To ensure the possibility of using production-line technology in the construction of gas pipelines with diameters of 1620 mm while maintaining the construction rates achieved and high work quality, a special series of machines and machinery must be developed. These machines include the following: a rotor trench excavator with a digging depth up to 3.0 m, a rotor trench backfiller with a capacity of 2500 cubic meters per hour, stands for bending pipe, pipe welding platforms, internal aligners, setups for automatic nonrotational welding, stands for preparing pipe edges, equipment for monitoring welded joints, gantry pipelayers, and special tooling equipment.

The factors indicated above decrease the competitiveness of 1620 mm pipelines in the most important construction regions, i.e., in the regions of Western Siberia. For this reason, it is proposed that during the 11th and 12th Five Year Plans, according to the emerging trend of development, the relative fraction of 1420-mm pipelines will increase from 10.9% in 1980 to 20% in 1985 and up to 25% in 1990.

2.2.2. Increase in Working Pressure of Main Gas Pipelines When the working pressure of a pipeline increases, the parameters of the gas transported and the technical characteristics of pipes and equipment change.

The technological parameters of the process and of the gas transported depend on pressure and temperature. We shall examine the changes in gas parameters for pressures at the beginning of the gas pipeline equal to 55, 75, 100, and 120 kg cm^{-2}; the pressure at the end of the pipeline is to be determined with a degree of compression equal to 1.5 and constitutes, respectively, 36.7, 50, 66.7, and 80 kg cm^{-2} (see Table V).

The temperature regime of gas pipelines with high capacity is determined by the conditions of heat exchange with the surrounding medium

Table V Application of Working Parameters of Gas Transportation and the Thermodynamic Properties of Transported Gas at Different Pressures

No.	Name of parameter	Initial pressure in the gas pipeline, P_i, kg/cm^{-2}			
		55	75	100	120
1.	Final pressure on the section, P_f, kg cm^{-2}	36.7	50	66.7	80
2.	Average pressure on the section, P_{av}, kg cm^{-2}	46.5	63.3	84.5	101.3
3.	Average (over the section) coefficient of compressibility of the gas, Z_{av}	0.896	0.864	0.828	0.802
4.	Cofficient of compressibility of the gas at the inlet to the compressor station, Z_{out}	0.912	0.882	0.849	0.826
5.	Specific heat capacity of the gas, C_p, kcal/kg^{-1} deg^{-1}	0.608	0.649	0.704	0.745
6.	Relative specific weight of the gas with respect to air, Δ	0.55	0.55	0.55	0.55
7.	Joule–Thompson coefficient, D_i, deg cm^{-2} kg^{-1}	0.46	0.43	0.39	0.37
8.	Heat transfer coefficient, K_{av}, kcal/m^{-2} h^{-1} deg^{-1}	1.5	1.5	1.5	1.5
9.	Degree of compression, ϵ	1.5	1.5	1.5	1.5
10.	Adiabatic index of natural gas, K	1.3	1.3	1.3	1.3

and by the decrease in temperature due to expansion of the gas, i.e., due to the Joule–Thompson effect. For gas pipelines with low capacity, the temperature of the gas decreases primarily (more than 60–80%) due to heat exchange with the surrounding medium, while the temperature decrease due to the Joule–Thompson effect is only 3–5°C. The change in temperature accompanying transportation of a large quantity of gas with an initial pressure of 55–120 kg cm^{-2} is shown in Tables V and VI. As is evident from these tables, for high initial temperatures $t_i = 55$°C, the decrease in temperature along the pipeline due to the heat exchange with the surrounding medium practically equals the decrease due to the Joule–Thompson effect. For an initial gas temperature of 15°C, the temperature decrease due to heat exchange with the surrounding medium decreases considerably (for example, for $t_i = 15$°C and $P_i = 120$ kg cm^{-2}, $t_{\text{exch}} =$ 0.76°C), while the decrease in temperature due to the Joule–Thompson effect increases (for $t_i = 15$°C and $P_i = 120$ kg cm^{-2}, $t_{\text{J–T}} = 14.77$°C). Calculations have shown that for transportation of cooled gas with an initial temperature of −2°C, the gas is not cooled due to heat exchange with the surrounding medium but rather is heated; for example, for $P_i = 120$ kg cm^{-2}, the gas is heated by 1.43°C, while the temperature decrease due to the Joule–Thompson effect is 17.17°C. This is a result of the large pressure drop along a 100 km section, which reaches 30–40 kg cm^{-2}.

The temperatures of the gas at the end of a 1420-mm pipeline initially pressurized at 75, 100, and 120 kg cm^{-2} without cooling the gas at compressor stations are practically equal and constitute 33–35°C with an initial gas temperature of $t_i = 55$°C. For an initial gas temperature of $t_i = 15$°C, the gas temperature at the end of the pipeline section for these pressures will be 2.36°C, 0.74°C, and −0.53°C, respectively; when the gas is cooled at the beginning of the pipeline to −2°C, these quantities will equal −10.69°C, −13.52°C, and −17.74°C.

As is evident from Table VI, the average gas temperature for all variants of the variation in pressure (with $t_i = 15$°C) may be assumed to be 10°C. The remaining working gas-transportation parameters are determined for methane with density $\rho = 0.668$ kg m^{-3} with an average gas temperature of $t_{\text{av}} = 10$°C. The magnitudes of these parameters are presented in Table VI as a function of the initial pressure in the gas pipeline.

The indicators of metal consumption in pipeline construction for different working pressures are calculated using Eq. (3) (see Table VII and Figs. 5 and 6).

The results of the calculation of pipeline capacity as a function of the initial working pressure using Eq. (4) are presented in Tables VI and VII

Table VI Decrease in the Temperature of the Gas Due to Heat Exchange with the Surrounding Medium and Due to the Joule–Thompson Effect for Different Initial Pressures

Parameters		$t_{\text{heat exch}}$ °C	t_{JT} °C	$\Delta t = t_{\text{heat exch}} + t_{JT}$, °C	t_k, °C
$P = 55$ kg/cm^{-2}	$t_i = 55$°C	19.45	6.84	26.29	28.71
	$t_i = 15$°C	2.61	8.68	11.29	3.7
$q = 24$ billion m/yr^{-1}	$t_i = -2$°C	−4.57	9.6	5.05	−7.03
	$t = 4$°C				
$P = 75$ kg/cm^{-2}	$t_i = 55$°C	12.98	8.65	21.63	33.37
	$t_i = 15$°C	1.87	10.78	12.64	2.36
$q = 32$ billion m/yr^{-1}	$t_i = -2$°C	−3.85	12.54	3.69	−10.69
	$t = 4$°C				
$P = 100$ kg/cm^{-2}	$t_i = 55$°C	10.45	10.67	21.12	33.9
	$t_i = 15$°C	0.93	13.23	14.26	0.74
$q = 45$ billion m/yr^{-1}	$t_i = -2$°C	−3.17	14.67	11.52	−13.52
	$t = 4$°C				
$P = 120$ kg/cm^{-2}	$t_i = 55$°C	7.77	11.98	19.75	35.25
	$t_i = 15$°C	0.76	14.77	15.53	−0.53
$q = 50$ billion m/yr^{-1}	$t_i = -2$°C	−1.43	17.17	15.74	−17.74
	$t = 4$°C				

Table VII Basic Characteristics of Gas Pipelines with Different Working Pressure

Working pressure P, kg/cm^{-2}	Coefficient of pressure increase	Thickness of pipe wall, δ, mm	Weight of 1 km of pipe, G, tons	Coefficient of increase of pipe weight	Coefficient of increase of capacity	Coefficient of variation of relative specific energy costs
55	0.733	11.0	382	0.707	0.73	1.03
75	1.000	15.6	540	1.000	1.00	1.00
100	1.333	21.6	744	1.377	1.33	0.96
120	1.600	26.3	920	1.703	1.58	0.94

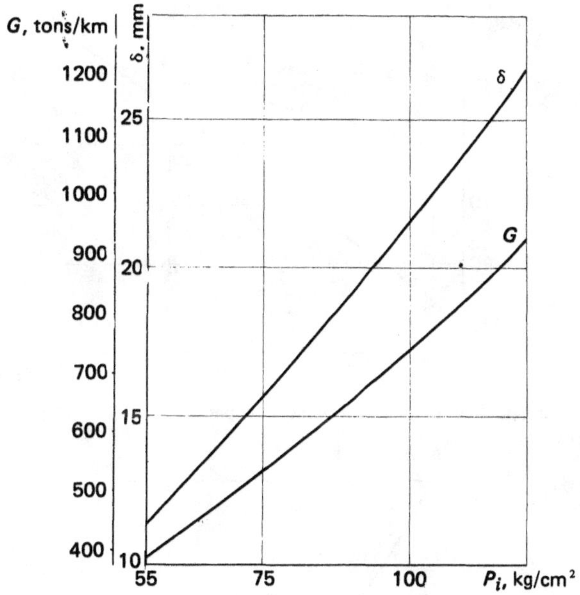

Figure 5. Dependence of wall thickness δ of the gas pipeline and weight of 1 km of pipe G on the working pressure.

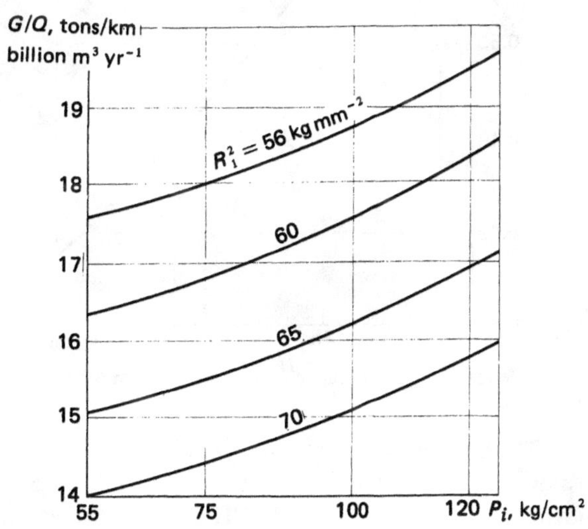

Figure 6. Change in the specific input of metal G/Q with increase in pressure and improvement of quality of metal.

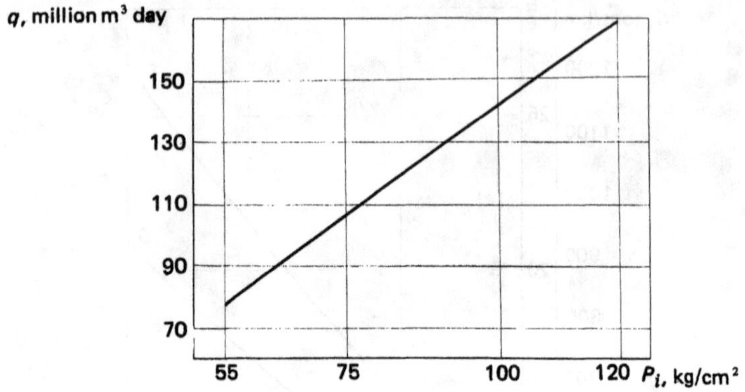

Figure 7. Dependence of the through-put capacity of 1420-mm gas pipeline on initial pressure.

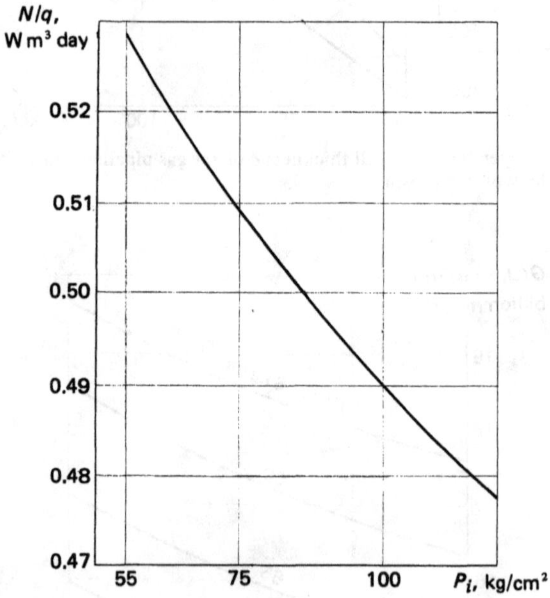

Figure 8. Dependence of specific capacity N/q of 1420-mm gas pipeline on the initial working pressure P_i.

and in Fig. 7. The change in the relative specific energy consumption for gas compression are presented in Table VII and in Figs. 8 and 9.

It is evident from the data in Table VII and Fig. 5 that for constant strength characteristics of the pipe metal, as the working pressure increases, there is a tendency for the thickness of the wall and the specific consumption of metal to increase.

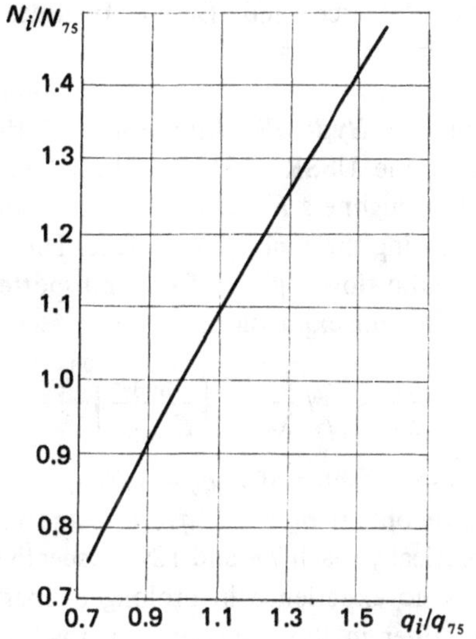

Figure 9. Dependence of the reduced working capacity on the reduced through-put capacity of gas pipeline.

For this reason, increasing the working pressures is ineffective if the strength characteristics of the pipe metal remain unchanged. It follows from here that the main condition under which the increase in working pressure becomes an efficient direction of technical progress is an improvement in the quality of the pipe metal. The latter eliminates the necessity of increasing the thickness of pipe walls with increasing working pressure. The dependence of the specific metal consumption on the working pressure for different strength characteristics of the pipe metal is presented in Fig. 6. When R_1^H changes from 60 kg cm^{-2} to 70 kg cm^{-2} with a working pressure of 75 kg cm^{-2}, the specific metal consumption by the gas pipeline decreases by 2.32 tons per kilometer per billion cubic meters. Thus an increase in the working pipeline pressure based on an improvement in the quality of pipe steel is very effective in mainline gas transportation.

Another direction for increasing the efficiency of gas pipeline transportation could appear due to the application of fundamentally new pipe designs, permitting the use of steel which does not contain expensive and scarce trace alloying elements with identical specific metal consumption due to an increase in the homogeneity and strength of thinsheet rolled steel [12]. One such design is the multilayer pipe design proposed by the E. O. Paton Institute of Electrical Welding of the Ukrainian SSR Academy of Sciences. The two-layer pipe design developed in the USSR at the

All-Union Scientific Research and Design Institute of Metallurgical Machine Building has an analogous goal.

2.2.3. Decrease in the Hydraulic Resistance of Main Gas Pipelines Studies performed in the USSR [13] and abroad [10] show that the equivalent haydraulic roughness $E_{\mathbf{hydr}}$ of pipelines can be decreased by a factor of 2–3 by coating the inner pipe surface. The increase in pipeline capacity with a quadratic flow regime of the transported medium can be estimated from the following expression:

$$\frac{q_2}{q_1} \approx \sqrt{\frac{\lambda_1}{\lambda_2}} \approx \left(\frac{E_{\mathbf{hydr2}}}{E_{\mathbf{hydr1}}}\right)^{0.1}.$$

If $E_{\mathbf{hydr}}$ decreases by a factor of 2, $q_2 = 1.07q_1$, and if $E_{\mathbf{hydr}}$ decreases by a factor of 3, we obtain $q_2 = 1.12q_1$, i.e., the increase in pipeline capacity could theoretically reach 7% and 12%, respectively.

However, there is no experience in prolonged operation of pipelines with such coatings either in this country or abroad and the results are based on experimental data obtained either under laboratory conditions or in compartively short test sections. For this reason, scientific-research and test-design work is being planned in the USSR to search for efficient polymer compositions based on readily available and inexpensive materials, as well as a technology for depositing such materials on interior pipe surfaces. Such surfaces must be stable to wear and should not peel off during long-term operation, and the accompanying increase in the cost of pipes should not exceed the increase in gas pipeline capacity. This measure can be implemented no earlier than the 12th Five-Year Plan.

Another means of decreasing the hydraulic resistance in pipelines could be the addition of small concentrations of polymer solutions, finely dispersed solid particles, etc., into the turbulent flow, which would decrease the haydraulic losses with the same velocities. However, these investigations are still in the laboratory stage.

2.2.4. Increase in the Number and Capacity of Compressor Stations In recent years, a number of arguments have been proposed in favor of increasing mainline capacity above the optimum value by increasing the number and capacity of compressor stations. These proposals are based on the fact that the cost of compressor stations in the overall cost of main-line construction does not exceed 30–40%, while the cost of the linear part of the pipeline exceeds 60–70%. We shall esitmate the efficiency of this measure. We shall first examine the energy intensiveness of gas transportation. It follows from Eq. (1) that when pipeline capacity increases from q_1 to q_2, the distance between compressor stations (other parameters remaining constant) must decrease from L_1 to L_2:

$$L_2 = L_1\left(\frac{q_1}{q_2}\right)^2. \tag{5}$$

In addition, according to Eq. (2), the capacity of each compressor station (newly constructed and operating stations) must increase from N_1 to N_2 and will constitute

$$N_2 = N_1\left(\frac{q_2}{q_1}\right). \tag{6}$$

Assuming now that the capacity of all compressor stations and the distances between them are identical, while the length of the pipeline equals 1, we obtain from Eq. (5) the number of compressor stations replacing n_1 before the increase with n_2 after the increase (after the capacity increases from q_1 to q_2):

$$n = \frac{l}{L_2} = \frac{l}{L_1}\left(\frac{q_2}{q_1}\right)^2 = n_1\left(\frac{q_2}{q_1}\right)^2. \tag{7}$$

Then, based on Eqs. (6) and (7), the total capacity of compressor stations on main gas pipelines will equal

$$n_2N_2 = n_1N_1\left(\frac{q_2}{q_1}\right)^3. \tag{8}$$

Thus, when the pipeline capacity increases from q_1 to q_2, the total working capacity of compressor stations must increase in proportion to the ratio of pipeline capacities raised to the third power. In addition, the surface area of the gas cooling apparatus at compressor stations must increase in proportion to the increase in the power consumed in compressing gas; Eq. (8) does not take into account the pressure losses in the connections in the compressor station, which increase in proportion to the number of compressor stations and in proportion to the square of the gas vélocity.

It should also be noted that the equipment in mainline compressor stations is subjected to very rapid wear and aging, so that during the operational lifetime of the gas pipeline (30–35 years), gas pumping aggregates must be replaced two to three times, which additionally complicates pipeline operation. Naturally, since in deriving Eq. (8) it was assumed that the distance between compressor stations and the gas flow rate through them are identical, the results obtained will be entirely valid for compressor stations with electrically driven plants or for compressor stations driven

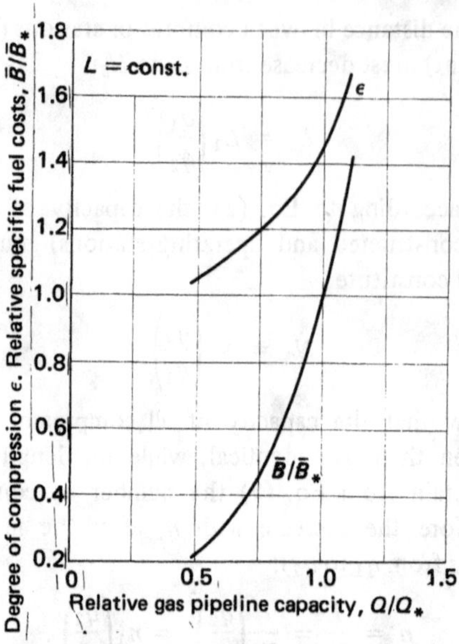

Figure 10. Relative specific fuel costs with different gas pipeline capacity and degree of compression at compressor stations (L = const.) (B_* and Q_* are the specific fuel costs and capacity of standard gas pipeline).

by heat engines, if the fuel gas, in this case, is extracted from other, neighboring gas pipelines.

For compressor stations using the gas transported along the same pipeline for fuel, the expression for the total power will be somewhat different and more complicated, but in this case the total power will be very nearly cubic (see Fig. 10).

As an example we shall now examine two variants of a system for transporting 200 billion cubic meters of gas per year (see Fig. 11) along 1420-mm pipelines pressurized at 100 kg cm^{-2}. In the first variant, gas is transported along five strings with each string having a capacity of 40 billion cubic meters of gas per year (the optimum pipeline capacity for fuel gas priced at 10 rubles per 1000 cubic meters). In this case, 30 compressor stations includes four working and three backup plants (scheme 4 + 2); at the end of the gas pipeline we obtain 175 billion cubic meters (35 billion cubic meters for each string) of commercial gas. In the second variant (see Fig. 11), the same volume of gas is transported over a four-string system with each string having a capacity of 50 billion cubic meters of gas per year. Then, the volume of commercial gas at the end of the system will be 164 billion cubic meters.

Technical-economic calculations have shown that the second variant

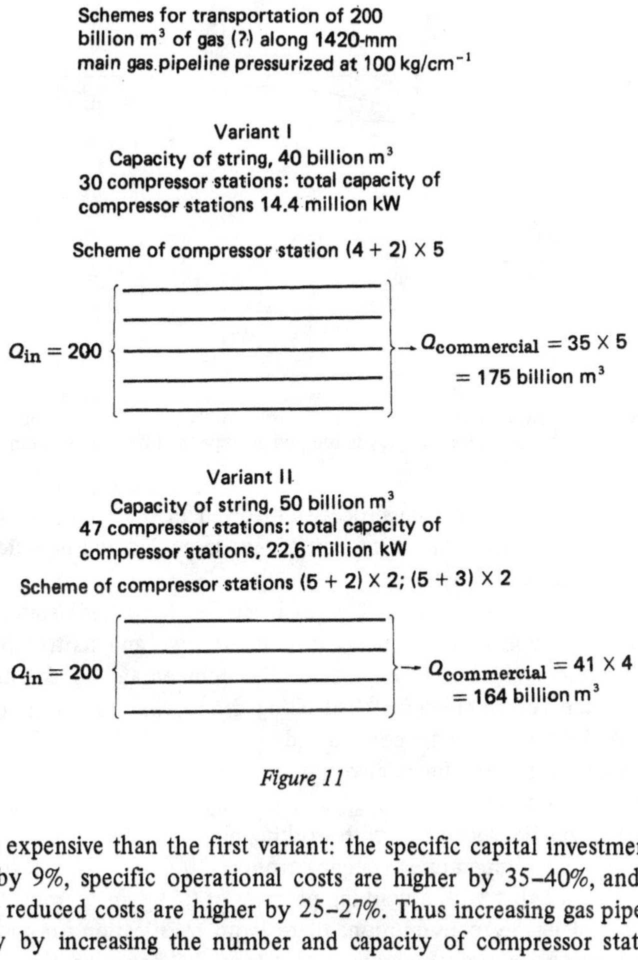

Figure 11

is more expensive than the first variant: the specific capital investment is higher by 9%, specific operational costs are higher by 35–40%, and the specific reduced costs are higher by 25–27%. Thus increasing gas pipeline capacity by increasing the number and capacity of compressor stations is economically unjustified.

2.3. *Nontraditional Methods for Transporting Gas*

2.3.1. Transportation of Cooled (to −70°C) and Liquified Gas It follows from expressions (1) and (2) that a simultaneous increase in the working pressure and cooling of the transported gas could have a large effect on the increase in capacity and decrease in the specific energy consumed for compression since both of these factors together also contribute more effectively to the decrease in the coefficient of compressibility of the gas.

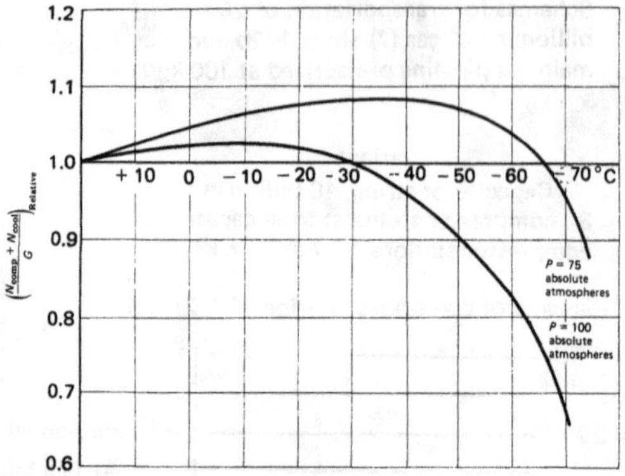

Figure 12. Power indicators of transportation of cooled gas as a function of the pressure and temperature. $N_{\mathbf{comp}}$ is the power expended on compression, $N_{\mathbf{cool}}$ is the power expended on cooling.

In this connection, the processes involved in the technical-economic indicators of transportation of natural gas in cooled and liquefied states were investigated in [8, 9].

The schemes for transporting gas in cooled (liquefied) states involve preliminary cooling or liquefaction of natural gas and its transportation along a thermally insulated pipeline. To compensate for the inflow of heat to the cooled (liquefied) gas along the pipeline route, intermediate stations for cooling gas are constructed every 400–600 km.

Figure 12 shows the relative energy indicators (relative energy consumption for compression $N_{\mathbf{comp}}$ and cooling of the transported medium $N_{\mathbf{cool}}$) of gas transportation with working pressures of 75 and 100 kg cm^{-2} and different temperatures (ranging from + 20°C to − 80°C). It follows from Fig. 12 that with a working pressure of 75 kg cm^{-2}, from the point of view of energy consumption, there is no justification for cooling the transported gas in the temperature range from 20°C to − 65°C.

For working pressures of 100 kg cm^{-2}, a decrease in the gas temperature below − 30°C decreases the consumption of energy for gas transportation. In long, high-capacity gas transportation systems (exceeding 2500 km in length and 50 billion cubic meters per year at a working pressure of 100–120 kg cm^{-2} and a temperature of minus 60–70°C), it is possible to decrease the specific energy consumption by 20–25%.

A decrease in gas temperature has a favorable effect on the specific consumption of metal by main gas pipelines. As the temperature of the transported gas decreases, the efficiency with which pipe metal is used in the mainline increases. At a gas temperature close to the critical value (around minus 70–75°C), the specific metal consumption by a transportation system constructed from 1420-mm pipes pressurized at 100 kg cm^{-2} decreases approximately two-fold compared with the present gas-transportation technology. Further decrease in temperature below the critical value (minus 82°C) and transformation of the gas into the liquefied state decrease the specific metal consumption by the gas pipeline by another factor of 1.5–2 due to the decrease in working pressure to 55 kg cm^{-2}.

The characteristics of low-temperature gas pipelines mentioned above and their low consumption of metal and energy as compared to the usual gas pipelines can, under definite conditions, also improve the economic indicators of long-range gas transportation.

To estimate the efficiency of low-temperature gas pipelines, calculations were performed at VNIIGAZ optimizing the technological parameters of gas transportation systems and the basic technical-economic indicators of variants of gas transportation at temperatures varying from + 20°C to – 120°C were compared.

A comparison of the transportation variants at temperatures of 20, – 30, – 70, and – 120°C indicates the following:

according to the total indicator, i.e., the specific reduced costs, the variants of gas transportation in the usual, cooled (to – 70°C), and liquefied states are practically identical;

the transportation variant with a temperature of – 30°C has the worst (compared to all the remaining variants) technical-economic indicators; the specific reduced costs for construction and operation of gas pipelines for this scheme of gas transportation are 10–12% higher than the variant with positive temperatures;

a decrease in gas temperature leads to considerable savings in pipe metal, constituting 42% for transportation of cooled (to – 70°C) gas and 72% for transportation of liquefied natural gas (LNG);

the variants in which gas is transported at a temperature of – 70°C has the lowest energy consumption.

It should be kept in mind, however, that the transition to a new transportation technology requires the solution of a number of scientific-technical problems.

Thus, for example, to realize the technology of transporting gas in the cooled state, it will be necessary to develop the following:

Steels and pipes, made of these steels for transporting gas at low temperatures with high strength and plastic characteristics, that are not more than 1.3–1.5 times more expensive than the usual pipes;

economic thermal insulating materials based on polyurethane;

centrifugal superchargers with a capacity of 60 and 10 000 kW at a working pressure of 100 kg cm^{-2} and a temperature of $-70°C$;

shutoff fixtures with diameters ranging from 500 to 1400 mm at a working temperature of $-80°C$ and pressure 100 kg cm^{-2};

shell-and-tube and spiral heat exchange apparatus for refrigeration systems with specific heat exchange surface not less than 10,000 square meters;

the technology and technical means for constructing thermally insulated 1420-mm pipelines for cooled gas at a working pressure of 100 atm.

The development and assimilation of series production of the entire complex of equipment for transporting gas cooled to $-70°C$ apparently requires at least ten years.

In realizing the technology for transporting liquefied natural gas (LNG), it will be necessary to solve the same problems, only instead of centrifugal superchargers it will be necessary to develop pumps for pumping LNG and turbine-driven compressor plants with large capacity (80–100 thousand kW) for compressing the coolant.

2.3.2. Combined Transport of LNG and Oil in the Frozen State The questions of combined transport of oil and liquefied natural gas along low-temperature pipelines were first raised and analyzed by Canadian specialists [14, 15] in connection with the assimilation of Arctic fields in Canada.

The fundamental scheme of combined transport of LNG and frozen oil is analogous to the scheme for transporting natural gas in the liquefied state. It involves preliminary liquefaction of natural gas, introducing the LNG and oil into a special mixer which yields a suspension of LNG and small frozen particles of oil, and pumping the gas–oil mixture with pumping stations along a thermally insulated pipeline. After-cooling of the transported mixture is provided for along the pipeline route. It should be noted that the technology of combined transportation of LNG and oil is extremely complex and can hardly be realized for long transportation systems.

In confirmation of what was said above, we shall present the basic technical-economic indicators of combined and separate transportation of LNG and oil over a distance of 3000 km. The specific indicators of combined transportation of LNG and oil, scaled to the volume of the transported gas, show 30% higher consumption of metal, 75% higher consumption of energy, and 50% higher specific reduced costs.

2.3.3. Transformation of Natural Gas into Methanol for Gas Transportation The production of methanol from natural gas is a complex and energy-intensive process.

The production of one ton of methanol raw material requires 999–1150 cubic meters of natural gas and 1160–1210 kWh of electrical energy.

The calorific value of methanol is 1.7–1.5 times lower than that of natural gas. For this reason, from the data presented above it follows that about 50% of the starting thermal energy of the gas is expended on the production of methanol. For comparison we note that in transporting natural gas over a distance of 3000 km over pipelines, about 12% of the gas is consumed, i.e., four times less gas than for the production of methanol. The overconsumption of energy indicated is not compensated by the insignificant economic effect, which is achieved in the linear part due to replacement of the gas pipeline by a methanol pipeline.

For the reasons indicated, this technology cannot be viewed as an alternative to pipeline transportation of gas.

2.3.4. Transportation of Natural Gas in the Form of Crystal Hydrates The essence of the technology of gas transportation in the form of crystal hydrates is as follows.

At the head end of pipelines, special installations are constructed for producing hydrates of natural gas.† Hydrates obtained in columns are packed in special containers, cooled to a temperature of minus 30–40°C and inserted into a pipeline. The hydrates are transported along the pipeline by pneumatic transportation of containerized loads. Compressed gas is used as the motive force. For this reason, compressor stations and special lock chambers for transferring containers from a low-pressure zone into a high-pressure zone are constructed along the pipeline. At locations where the gas is utilized, the hydrates are reconverted into the gaseous state yielding commercial gas and water. An additional pipeline string is required for returning the empty containers.

The method described is inefficient and involves a considerable increase in transport operations and energy consumption. It is enough to note that 6 to 7 kg of ballasting water must be transported for each kg of natural gas.

Calculations show that the transportation of 30 billion cubic meters of natural gas per year over a distance of 3000 km in the form of crystal hydrates requires about 5.0 million kW of compression equipment, which

† Hydrates represent a physical compound of gas with water in a ratio of 1 to 6–7, i.e., for each molecule of gas (methane), there are 6–7 molecules of water.

is almost three times more than for the usual gas pipelines with the same capacity and length.

The metal consumption in the transportation system remains at the previous level, while the capital investment and operational expenses increase by a factor of 1.5 and 2.2, respectively.

Thus the analysis performed above shows that in the immediate future (10–15 years) none of the nontraditional methods for transporting gas examined above are competitive with the traditional method of pipeline transportation in the gaseous state. For this reason, new methods must be sought for increasing the mainline capacity within the framework of existing or close to existing technology.

One such path, as demonstrated previously, is to increase the working pressure in main gas pipelines. In going from a working pressure of 75 kg cm^{-2} to 100 kg cm^{-2}, the pipeline capacity increases by 30–35%, which solves the problem of increasing the capacity of gas pipelines, but since in this case metal consumption increases considerably, the specific metal and capital costs and the reduced costs practically do not decrease, while the fuel-energy costs decrease by 4–5%. For this reason, it is useful to make the transition to this pressure level. It should be noted that the construction of such a gas pipeline is a complicated scientific-technical problem, and there is no experience in world practice of constructing and operating gas pipelines with such diameters pressurized at 100 kg cm^{-2}. The maximum diameter of gas pipelines presently constructed abroad does not exceed 1200 mm.

For this reason, a complex special purpose program for solving this problem has been instituted to create and develop a gas-transportation technology based on optimum parameters of high-pressure gas pipelines (optimum capacity, optimum scheme and degree of compression in compressor stations, optimum compressor station spacing, optimum level of gas cooling in different climatic zones), taking into account the trend in the prices of pipe, gas pumping aggregates, and other technological equipment, as well as the price of fuel gas.

To construct main gas pipelines pressurized at 100 kg cm^{-2}, it is necessary to develop and assimilate the production of an entire complex of equipment (pipes, connecting fixtures, shutoff-regulating fittings, dust traps, filters-separators, gas-pumping aggragates as well as new construction machines and mechanisms). This program is currently being successfully implemented.

It should be noted that as the working pressure increases up to 100 kg cm^{-2}, the seriousness of the consequences of accidental breaks of such pipelines increase sharply. For this reason, it is proposed that intially these gas pipelines be restricted to sparsely populated regions of Western Siberia. In future, after experience has been accumulated in designing, constructing, and operating such pipelines and the reliability of gas pipelines pressurized at 100 kg cm^{-2} increases, such pipelines could be constructed in populated regions as well.

3. Energy Conserving Technological Processes and Solutions to Technical Problems

3.1. Design of Main Gas Pipelines for Optimum Capacity and Optimum Operation

The energy intensiveness of gas transportation is affected most by pipeline capacity. As shown in Section 2.2.4, energy consumption is proportional to the third power of the capacity ($\sim Q^3$).

Investigations at VNIIGAZ [16–18] and other design and research institutes have established that each standard size of pipe (depending on the costs of constructing and operating main gas pipelines, as well as the ratio of the costs for the linear part and for compressor stations) corresponds to an optimum throughput, satisfying the condition of minimum reduced costs. In addition, near the optimum point, the unit reduced costs vary (with an increase in the throughput of gas pipelines by 2–5%) insignificantly. Further increase in the through-put of gas pipelines leads to an sharp increase in energy consumption (fuel consumption) for gas transportation. Thus, for example, an increase in the throughput of a 1420-mm pipeline pressurized at 75 kg cm^{-2} from 31 to 36 billion cubic meters (by 16%) will increase the fuel costs by 40–50%.

A decrease in pipeline capacity is accompanied by the opposite process: decrease of energy consumption for gas transportation. For this reason, due to the increased cost of energy resources, a number of arguments have recently been proposed in favor of lowering the capacity of gas pipelines in order to decrease energy consumption. We shall examine this problem in greater detail.

Gas pipeline capacity can be decreased in three ways: (1) by increasing the distance between compressor stations with the degree of compression remaining unchanged; (2) by decreasing the degree of compression with the distance between compressor stations remaining constant; and (3) by a conbination of the first two methods, i.e., increasing the spacing and decreasing the degree of compression in compressor stations. In the first case, a decrease in pipeline capacity by 25% decreases, according to Eq. (8)

in Section 2.2.4, the overall consumption of energy for gas transportation by 55–57%, while the specific energy consumption (referred to unit pipeline capacity) decreases by 44%. In the second case, the overall energy consumption, according to Fig. 10, decreases by 59–60%, i.e., somewhat more than in the first case. For this reason, the second method of decreasing energy consumption is preferable. The indicators for the third method will fall between the first and second methods. However, in all cases, the unit metal consumption will increase (inversely proportional to pipeline capacity). For this reason, the basic technological parameters of gas pipelines (including the capacity) are chosen based on optimization calculations, in which a reasonable compromise is achieved between the different technical-economic indicators of gas transportation (between specific energy-metal and capital costs) taking into account the limits on the material, energy, and labor resources. Therefore, pipeline capacity cannot be selected starting from a single gas transportation indicator only. In planning and designing main gas pipelines, provision for their optimal capacities should be kept in mind.

Measures optimizing the operation of gas pipelines can also greatly reduce energy consumption. Maintaining optimal regimes, according to estimates by experts, will save up to 5% of the total gas consumed for internal needs of compressor stations along main gas pipelines.

Optimum regimes with a given capacity must be maintained over the course of pipeline operation. It was shown in [19, 20] that for pipeline capacity exceeding one-half the planned value, minimum energy consumption for gas transportation corresponds to a regime in which maximum working (design) pressure is maintained at the compressor station outlet. For pipeline capacity lower than one-half the planned value, a problem arises in redistributing the load between compressor stations. For these purposes, computer optimization methods have been developed and introduced in the USSR [19, 21]. A complex of programs called Optimum has been developed and is used by the gas industry in this country as part of the automated dispatcher control system of the unified gas supply system of the USSR. The purpose of this program is to find the optimum variants according to such criteria as maximum throughput of the pipeline and minimum energy consumption and, in addition, the latter indicator can be expressed both in percent and in natural units: power consumed by compressor stations, consumption of fuel gas, and total energy consumption of gas transportation.

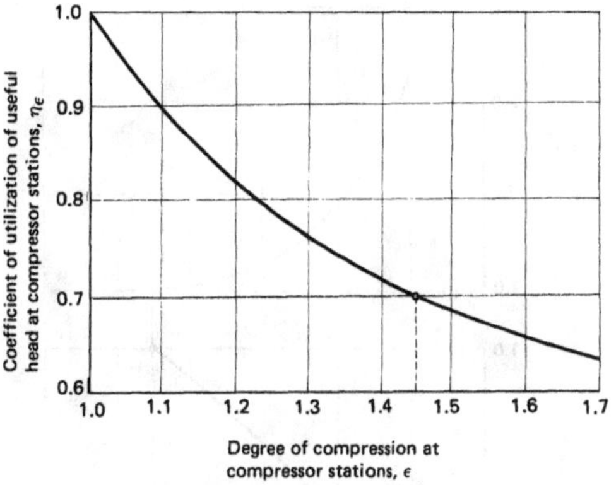

Figure 13. Dependence of the coefficient of utilization of the useful head on the degree of compression at compressor stations.

3.2. *Low-Head Gas Transportation Technology*

The prospects of using low-head technology for transporting gas along main gas pipelines was justified in investigations performed at VNIIGAZ (S. N. Sinitsyn, E. V. Leont'ev, and I. V. Vartsev) at the end of the 1960s. The physical prerequisite, ensuring the efficiency of the low-head technology, is a decrease in the head losses in gas pipelines due to an increase in the average pressure of the transported gas accompanying a decrease in the distance between compressor stations and the degree of compression.

The degree of compression at compressor stations determines the efficiency with which the head transferred to the pipeline (H) is used. The utilization efficiency of the useful head at compressor stations can be estimated, according to S. N. Sinitsyn and E. V. Leont'ev, by the coefficient

$$\eta_e = \lim_{\epsilon=1} \Sigma \overline{H}$$

where $\overline{\Sigma H} = \Sigma H/(\Sigma H_0)$ is the relative total polytropic head of the compressor station; (ΣH_0) is the total head of the compressor station with a degree of compression $\xi_0 = 1.45$.

The dependence of the coefficient of utilization of the effective head of the compressor station on the degree of compression is presented in Fig. 13. The utilization efficiency of the effective head on operating gas pipelines (with a degree of compression $\xi = 1.45–1.5$) corresponds mainly to $\eta_\xi = 0.65–0.7$. One way to improve the gas transportation technology and to decrease energy consumption is to decrease the degree of expansion of the gas along the linear sections, i.e., by introduction of low-head

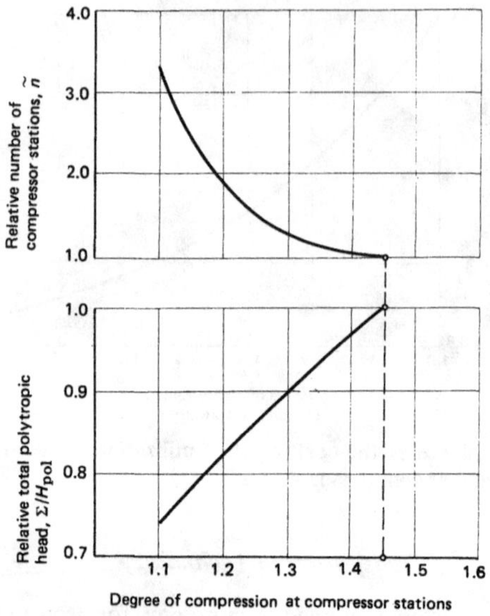

Figure 14. Relative change in the number of compressor stations and the total polytropic head as a function of the degree of compression at compressor stations.

compressor stations (with a fixed working pressure without decreasing pipeline capacity). As shown in Fig. 14, a decrease in head losses on sections (and, therefore, decrease in compression and total power, necessary for compression) requires a relative increase in the number of compressor stations along the gas pipeline accompanying a decrease in the distance between them. It is evident from Fig. 14 that a decrease in the degree of compression at compressor stations from the average attained level 1.45 to 1.25 (degree of compression of single-step supercharges) decreases the total polytropic head of compressor stations by approximately 14%. The specific consumption of working capacity (per 1000 cubic meters of gas) will decrease by the same amount. The number of compressor stations on the gas pipeline will increase approximately by a factor of 1.5. Figure 15 shows [7] the dependence of the degree of compression and the relative unit fuel costs $\bar{B}/\bar{B}_*$ on the relative spacing of the compressor stations L/L_* with constant capacity (the parameters $\bar{B}_*$ and L_* refers to a gas pipeline with the degree of compression equal to 1.45).

It is evident from Fig. 15 that a decrease in the degree of compression from 1.45 to 1.35 (degree of compression of low-head removeable flow-through part of perspective centrifugal superchargers), accompanying a

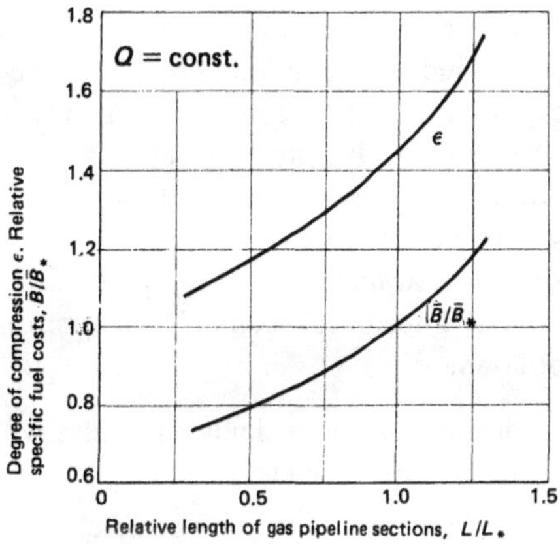

Figure 15. Relative specific fuel costs with different degree of compression and spacing of compressor stations ($Q = \text{const.}$) (B_* and L_* are the specific fuel costs and length of sections of standard gas pipeline).

decrease in the spacing of compressor stations by 14%, permits decreasing the specific energy costs by 7%, while maintaining the pipeline capacity at the previous level (as with $\epsilon = 1.45$). Introduction of low-head technology without reduction of the compressor station spacing requires looping of sections of the gas pipeline. Calculations show that when the degree of compression is decreased to 1.35, it is necessary to lay looping pipelines with a length equal to 14% of the length of the section with diameter equal to the diameter of the main pipe. The specific fuel costs will decrease by a much larger percentage: by 20%. The choice of the most efficient variant of low-head technology for transporting gas must be justified in each specific case: the optimum values of the spacing and degree of compression of compressor stations correspond to the optimum capacity. Often, during planning (taking into account the uncertainty of technical-economic information), the pipeline capacity is assumed to be somewhat higher than the optimum value, due to the increase in the degree of compression of compressor stations. As shown in Fig. 12, this increase in capacity leads to an increase in the specific energy consumption for gas transportation. As shown in Fig. 15, an increase in the degree of compression from 1.45 to 1.55 (degree of compression of high-head modifications of perspective superchargers) with the compressor station spacing remaining constant increases the capacity of the gas pipeline by up to 5% and leads to an increase in the specific fuel costs by 15–17%.

The prospects for introducing the low-head technology is viewed by VNIIGAZ as a planned and gradual process of decreasing the average degree of compression of compressor stations as the UGSS increases in size. In addition, it should be kept in mind that both the existing and low-head technologies require modifications of the removeable flow-through part of the superchargers, which ensure efficient loading of the drive under conditions when there is a spread in the length of pipeline sections and gas is extracted along the pipeline route.

Introduction of low-head technology has up to now been held up by a number of technical and technical-economic factors, of which the most important are as follows:

laying new gas pipelines along assimilated routes with connection of new compressor plants to existing pads;

the large fraction of the cost of the infrastructure in total costs of constructing and operating gas pipelines.

At the beginning of the 1980s, the problem of a low head was examined again in connection with changing conditions and characteristics of the present stage of development of the UGSS.

We can list a number of reasons for examining this problem again:

change in the starting technical-economic information on pipes, gas pumping aggregates, fuel gas, etc.;

possibility of introducing low-head compressor stations along gas-pipeline routes newly opened up during the 11th Five-Year Plan;

increase in the capacities of the construction-assembly organizations and improvement of methods for constructing objects along main gas pipelines;

taking into account the conditions for specific planning of gas pipelines (pressure losses in ties with compressor stations, possible standard sizes of equipment used, etc.).

From the investigations performed, we can draw the following conclusions.

Low-head variants of gas transportation along gas pipelines pressurized at 75 kg cm^{-2} can be realized based on plants with unit capacity no greater than 16,000 kW.

When equipping compressor stations with 10,000 kW plants, the low-head technology permits decreasing the installed capacity by 12–15% (the higher value refers to compressor stations with a pressure drop of 1.5 kg cm^{-2} in the compressor station connections and the lower value refers to a pressure drop of 2.5 kg cm^{-2}).

In equipping compressor stations with 16,000 kW plants, the corresponding decrease in the working capacity is 5 and 10%.

The larger decreases in installed capacity in the variant with 10,000 kW plants are explained by the additional effect of different conditions for backing up the gas pumping plant in the low-head and traditional variants. The decrease in the installed capacity obtained for compressor stations with plants of 16,000 kW reflects the pure influence of the low-head technology.

The technical-economic efficiency of introducing low-head technology is primarily determined by the percentage of reduced expenditures which do not depend on the capacity of the compressor station and expenditures on the infrastructure.

In equipping compressor stations with 10,000 kW plants, the low-head gas transportation technology provides for a decrease of specific reduced costs by an amount varying from 1 to 4%.

In equipping compressor stations with 16,000 kW plants, the traditional and low-head gas transportation technologies are characterized by approximately the same specific reduced costs.

3.3. *Increase in the Hydraulic Efficiency of Main Gas Pipelines*

The unsatisfactory quality of pipeline construction and the unsatisfactory state of the pipeline upon release (construction debris and water remaining in the pipeline), as well as unsatisfactory preparation of gas for long-distance transportaion increase the hydraulic resistance and, therefore, decrease the hydraulic efficiency of pipelines as well, sometimes down to 0.90–0.92 (ratio of the actual pipeline capacity to the possible capacity under the same pressure differentials). An increase in the relative fraction of the gas from northern fields, where installations for drying gas down to the temperature of the dew point of minus 20–25°C will be built, as well as an improvement in the quality of gas preparation in fields in Central Asia, periodic cleaning of gas pipelines by passing mechanical pistons through the pipeline, improvement of the quality of construction and the state of the pipeline upon release for operation will increase the hydraulic efficiency of main gas pipelines up to 95%. The complex of measures indicated will permit either increasing the pipeline capacity by 5% without increasing fuel costs or decreasing fuel costs without increasing pipeline capacity. We shall examine in greater detail the problems of decreasing energy consumption for gas transportation while increasing the hydraulic efficiency of gas pipelines.

We shall assume that the coefficient of hydraulic efficiency of gas pipelines ($E_{\mathbf{hydr}}$) is 90%. To maintain the maximum planned pipeline capacity ($E_{\mathbf{hydr}} = 100\%$), it is necessary either to construct additional looping (which is unlikely) or, in the presence of available backup capacity at compressor stations along mainlines, to increase the degree of compression of compressor stations. In the latter case, naturally, energy consumption will increase. We shall estimate this increase. We shall first determine the increase in the degree of compression ϵ using Eq. (1) in Section 2.1. In so doing, we shall assume that the planned working pressure levels (the most favorable regime) are maintained at the compressor station outlets; the degree of compression is increased due to the decrease in pressure at the compressor station inlet. We shall assume that for a coefficient for hydraulic efficiency equal to 1, the degree of compression is 1.45. We shall determine the degree of compression required so that the pipeline capacity is 100% with a coefficent of efficiency equal to 0.90.

The calculations showed that the degree of compression sought equals 1.68. In this case, the fuel (energy) consumption increases approximately by a factor of 1.30–1.37. Thus maintaining the hydraulic efficiency of gas pipelines at a high level represents a large reserve for decreasing the energy intensiveness of mainline gas transportation.

To fulfull, in practice, the measures required to increase the hydraulic efficiency, startup and piston inlet chambers, as well as scrubbing and calibrational pistons will be installed in all newly planned and operational gas pipelines with diameters of 1020–1420 mm; during operation, the gas pipelines will be cleaned with optimum periodicity. Realization of the measures listed above will permit maintaining the hydraulic efficiency of gas pipelines at a level of at least 95%.

3.4. Transition to Parallel Operation of Gas Pumping Aggregates with Full-Head Centrifugal Superchargers (CFS)

New types of gas pumping plants with full-head centrifugal superchargers have been developed and will be introduced during the 11th Five-Year Plan.

Parallel connection of gas pumping plants with full-head superchargers, compared with a parallel-series scheme of compressor stations, has a number of advantages, including lower fuel-energy consumption [4]. The latter is achieved due to the following factors:

operation of excess plant capacity is eliminated due to the possibility of switching on an odd number of machines;

pressure losses in the compressor station ties are decreased due to the simplified construction of the station;

the economic efficiency of operating gas pumping aggregates in capacity-regulated regimes increases.

Excess gas pumping aggregates can be eliminated at compressor stations connected to sections with low hydraulic resistance, as well as at all other compressor stations operating at partial capacity.

The possibility of operating an odd number of gas pumping plants makes it possible to decrease the fuel-energy consumption by 13–25% (depending on the number of plants at the station).

Taking into account the fraction of compressor stations connected with low-resistance sections and the relative duration of production regimes which permit shutting off one gas pumping plant (transition to an odd number of operating machines), the mean-weighted value of the fuel-energy costs is 2–3%.

The decrease in the fuel-energy costs, related to the decrease in pressure losses in the ties at the compressor station, is of the same order of magnitude. Thus the decrease in fuel-energy consumption with parallel connection of superchargers amounts to 4–6%.

3.5. Use of Replaceable Flow-Through Parts in Working Regimes

Fluctuations of the planned distances between compressor stations and of the ratio of the transitional and line flow rates give rise to a spread in the degree of compression at compressor stations. The statistical distribution of compressor stations with respect to the degree of compression, found in investigations conducted by VNIIGAZ [5], is shown in Fig. 16. It follows from Fig. 16 that the range of planned degrees of compression varies from 1.25 to 1.65; the center of the distribution (average degree of compression) $\xi_0 = 1.45$. The area under any section of the distribution curve is equivalent to the relative fraction of compressor stations (percent) located in the corresponding range of degrees of compression.

The use of a replaceable flow-through part in modifications of superchargers with respect to the degree of compression and pressure makes possible complete loading of installed aggregates in the entire range of technological parameters of line and head compressor stations. As the loading on of the aggregates increases, aside from a decrease in their total number, fuel is saved due to the higher effective efficiency of the drive.

The results of calculations of the relative decrease in fuel costs due to the use of the degree-of-compression modifications of CFS on line compressor stations are presented in Table VIII. These calculations were performed under the following conditions:

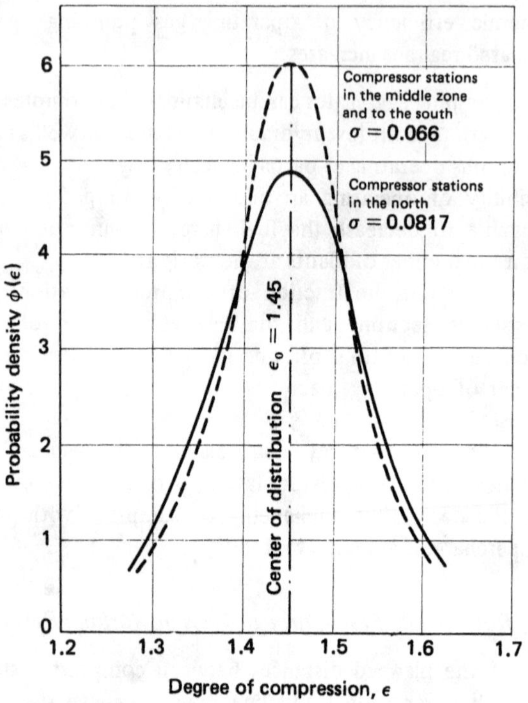

Figure 16. Distribution of compressor stations situated in different climatic zones on the degree of compression.

Table VIII Decrease in Fuel Costs with the Use of Replaceable Flow-Through Part in Compressor Stations Along the Pipeline

Compression range	1.55–1.46	1.46–1.39	1.39–1.32	1.32–1.26
Relative number of compressor stations	0.17	0.37	0.33	0.13
Relative load with flow-through part	1	1	0.79	0.64
Relative cost of fuel with flow-through part	1	1	0.94	0.87

identical polytropic efficiency of compression in the variants compared;

use of CFS with a degree of compression equal to 1.44 in the compared and reference variants;

Table IX Relative Decrease in the Cost of Fuel With the Use of Modifications of Centrifugal Superchargers According to the Working Pressure on the Head Compressor Stations

Working pressure of step, kg/cm^{-2}	76	56	40	29
Relative capacity of head compressor stations	0.25	0.25	0.25	0.25
Relative load without flow-through part	1	0.74	0.52	0.38
Relative cost of fuel with flow-through part	1	0.92	0.81	0.70

normal distribution of compressor stations over the degree of compression.

On the average, the fuel savings due to the use of replaceable flow-through parts on line compressor stations will amount to 4%.

The calculation of the relative decrease in fuel consumption accompanying the use of working-pressure modifications of CFS in head compressor stations was performed for the following conditions:

identical polytropic compression efficiency in the variants compared;

use of CFS at a pressure of 56 kg cm^{-2} in the compared and reference variants;

uniform distribution of capacity over degrees of compression at head compressor stations.

The results of the calculation are presented in Table IX.

On the average, the fuel savings due to the use of working pressure modifications will amount to 14%. Starting from the ratio of the capacities of line and booster compressor stations, the average percent savings in fuel will equal $\delta = 5\%$.

For an average annual increase in the amount of transported gas equal to $\Delta Q = 40$ billion cubic meters per year and fuel costs amounting to $\xi = 15\%$, during the 11th Five-Year Plan, the average annual fuel savings will be $\Delta G_T = 5.5 \Delta Q \delta \xi = 5 \times 5.40 \times 0.05 \times 0.15 = 1.65$ billion cubic meters per year.

3.6. Use of Input Control Devices (ICD) for Regulating the Capacity of CFS

The main reason for the lower efficiency of electrically driven gas pumping plants in the absence of turnaround control; control by throttling or

Table X Expected Power Consumptions with Regulation Using Input Control Devices (ICD)

Relative capacity, Q	0.8	0.85	0.9	0.95	1.0
Relative capacity without regulation, N	0.92	0.95	0.97	0.99	1.0
Relative capacity with regulation by ICD, N_r	0.68	0.75	0.82	0.90	1.02

by-passing the part of the transported gas from the supercharger line into the intake line leads to an average overconsumption of electricity by 5–9%. It is proposed that during the 11th Five-Year Plan a new class of turn-around controllable electrically driven gas pumping plants with higher capacity will be produced.

The installation of input control devices an efficient method for regulating the capacity of compressor stations with series produced and currently used gas pumping plants with electric drive [22].

Table X presents the dependences of the relative power consumption for capacity regulation at compressor stations with and without ICD for the operational range of electrically driven CFS.

In the period from 1981 to 1985, according to calculations performed at VNIIGAZ, the gas industry will require pumping plants with electrical drive power of 4 and 12.5 thousand kW totalling 3 755 000 kW. For the annual installation of electrically driven compressor stations with a total power of 716 000 kW, the installation of ICD will decrease the consumption of electricity with a nonuniformity factor of the gas pipeline equal to 0.92 by 36 million kWh and natural gas consumption by 2 million cubic meters per year. The savings of reference fuel will amount to 13 thousand tons per year.

3.7. *Staged Introduction of Capacity of Compressor Stations under Construction* [23]

The period during which gas pipeline capacity under construction is put into operation is characterized by low gas transportation efficiency. This is explained by the following factors:

higher pressure losses on gas pipeline sections in connection with decreased working pressure;

lower efficiency of the drive and of the superchargers operating in unplanned regimes;

suboptimal number of operating gas pumping aggregates, related to restrictions on the scheme for switching on plants, permitted by the technical connections;

overconsumption of power used in antipumping by-pass regimes.

The possibility of increasing the efficiency of gas pipeline operation during the startup period represents a large reserve of energy resources. This is explained by the large spread in startup regimes on pipelines under construction.

At the present time, up to 50% of the total volume of gas transported in the USSR is transported along such systems.

The following complex of measures is designed to decrease fuel-energy consumption by gas pipelines operating in the startup regimes:

improvement of the methods used in constructing compressor stations for the ultimate (future) purpose of allowing for simultaneous placement of compressor stations and the linear part of pipelines into operation;

use of a replaceable flow-through parts for CFS and input control devices to increase the working pressure in the gas pipeline, increase the economic efficiency of gas pumping aggregates, and eliminate the antipumping by-pass.

use of flexible technological connections of compressor plants in order to provide an optimum scheme for switching on the gas pumping aggregates.

It should be noted that the introduction of staged construction of compressor stations is entirely realistic right now. The essence of this principle lies in the following. All equipment and construction of compressor stations must be separated into startup complexes corresponding to the order of construction. The startup complex consists of a system comprising the basic and auxiliary equipment, necessary and sufficient for the period of gas pipeline operation under study. For compressor stations with up to four gas pumping plants, two startup complexes must be included. For a large number of gas pumping plants, the number of startup complexes per compressor station could equal two or three.

The efficiency of staged introduction of compressor station capacity increases if the replaceable flow-through part of centrifugal superchargers, corresponding to the gas pipeline regime at the given stage of its development, is used. The replaceable flow-through part could consist of modifications to the planned operational regimes (degree of pressure increase 1.35, 1.45, 1.55), as well as boosting modifications at low working pressures.

As an example, Table XI presents the program and efficiency of two-stage construction of compressor stations on a gas pipeline with 16 stations, equipped with eight gas pumping aggregates with full-head superchargers. The effectiveness of the measures indicated depends on the conditions of the specific gas pipeline (length, standard size of compressor station, type of CFS: full-head or not, etc.). The measures listed make it possible to decrease the fuel-energy costs by 20–40% during the startup period.

3.8. *Increase in the Efficiency of Equipment at Compressor Stations*

One of the key questions concerning conservation of fuel gas in compressor stations is the large increase in the average efficiency of the power drive stock, primarily, gas-turbine drive, which has the largest relative significance in the industry.

To satisfy the needs of mainline gas transportation, a program for developing new high-efficiency gas pumping plants is planned and in the process of being realized.

An increase in the average efficiency of the stock of gas pumping aggregates is to be achieved, primarily, by massive introduction (beginning with the 12th Five-Year Plan) of 16- and 25-MW plants with an efficiency of 29–30% and planned reliability indicators at the level of the best analogs.

The GTN-16 and GTN-25 block – no-basement automated plants with a capacity of 16 and 25 MW, respecitively, are intended for use on 1420-mm gas pipelines pressurized at 75 and 100 kg cm^{-2}. They are intended to work in different climatic regions with installation in individual easily-erected buildings. In combination with other technological equipment at the compressor station, they will increase the level of industrialization, decrease the volume of construction-assembly work, and decrease the time required for construction of compressor stations. Together with the GPA-Ts-16 transport-type plant with a capacity of 16 MW, the plants indicated will apparently provide the foundation for new capacity in the period 1985–2000.

The need to decrease the periods of time required for operational-assembly work, to decrease the capital investment in compressor stations, and to make the transition to unit and aggregate maintenance of gas-pumping aggregates led to widespread introduction of gas-pumping aggregates with gas-turbine engines used in transportation into the gas industry. At the beginning of the 10th Five-Year Plan, only two types of such aggregates were used in the industry: GPA-Ts-6.3 and GPA-10. During the

Table XI **Program and Efficiency of Two-Stage Construction of Compressor Stations Along a Single Gas Pipeline with 16 Compressor Stations Equipped with 6 Working and 2 Reserve Gas-Pumping Aggregates (Standard Size 6 + 2)**

Stages in construction of compressor stations	Number of compressor stations introduced at the given stage	Number of construction stage	Number of gas-pumping aggregates introduced as the given stage			Degree of pressure increase with modification of flow-through part	% assimilation of planned capacity of gas pipeline	Reduction of fuel-energy cost (%) compared with the method of construction with single-stage introduction of compressor stations	
			Total	Working	Reserve			Staged	Total
1	8	1	4	3	1	1.35	65	16	25
		2	4	3	1	1.55	78		
2	8	1	4	3	1	1.35	90	9	
		2	4	3	1	1.45	100		

11th Five-Year Plan, there will be widespread introduction of GPA-Ts-16 plants with superchargers at pressures of 75 and 100 kg cm^{-2}, and GPA-Ts-25 plants will be introduced in the 12th and 13th Five-Year Plans.

The GPA-Ts-16 aggregate has a block no-basement construction and it is automated. It is intended for operation in different climatic zones; it is placed in a factory-built container or in an easily erectable individual building. It provides for an increase in industrialization, decrease in the volume of construction-assembly work, and decrease in the construction time of compressor stations. The presence of replaceable modifications of the flow-through part permits satisfying the different technological requirements of the line and booster compressor stations. The system for technical maintenance and repair decreases requirements on the production repair stations, and decreases labor and repair time due to complex factory repair at Minaviaporm factories.

The GPA-Ts-10 gas-pumping plant is being developed due to forcing of the parameters of the thermodynamic cycle of the base NK-12 driving engine, which yields an effective power of 10 MW with an efficiency of 26%.

The production of these plants will be supported by a corresponding increase in the production of the GPA-Ts-6.3 plant, developed on the same foundation and having a power of 6.3 MW with an efficiency of 22%.

An important additional advantage of the development of the GPA-Ts-10 plant is the possibility of using newly produced engines to replace the exhausted GPA-Ts-6.3 engines. In this case, the parameters and remaining equipment in the compressor station, with the exception of the flow-through part of the superchargers, does not have to be replaced. As a result of the increase in the efficiency of the drive, such reconstruction will save on the order of 2.7 million cubic meters of fuel gas per year at each reconstructed plant.

The high parameters of the base aircraft engine, used to develop the GPA-Ts-25 plant, will permit, when it is converted, developing the most economic domestic plant of this type with an efficiency of 32–33%. The detailed advantages of the GPA-Ts-25 plant are analogous to those of the GPA-Ts-16 plant. Work on developing the GPA-Ts-25 plants will apparently be expanded during the 12th Five-Year Plan.

By the beginning of the period examined, a significant fraction of the gas-turbine stock (of the order of 12–15%) will have an operational lifetime exceeding or close to the planned lifetime (GT-700-4, GT-700-5, GTK-5, and GT-750-6).

These plants require that a complex of measures be implemented to restore the technical state. Plants having the worst power, efficiency, and reliability indicators are subject to being written off and replaced.

The requirement for simplifying the GTI schemes, following from the need to simplify construction-assembly operations, conflict with the need to increase the economic efficiency of GTI, dictating increased complexity of GTI schemes due to the introduction of regeneration. On this level, the most acceptable method for increasing the economic efficiency and simultaneously decreasing the unit mass is to increase the initial parameters of the working processes in GPA (temperature, pressure, high rates).

The problem of increasing the economic efficiency of prospective gas pumping aggregates can be solved due to the development and introduction of highly economical combined gas pumping aggregates: with steam-gas cycle, turbopiston drive, etc.

The scales for application of electrically driven GPA on prospective gas pipelines are determined by the following considerations.

Electrically driven gas pumping plants have a number of advantages over gas-turbine plants:

lower capital investment in compressor stations by 35–40%;

lower operational expenses (without the energy component) by 20%;

higher reliability of gas pumping plants;

lower fire and explosion hazard.

In the foreseeable future up to the year 2000, according to the present trend, electrical drive could be more widely used in connection with the development of nuclear power, which will decrease the consumption of gas for the internal needs of gas pipelines. According to calculations performed by VNIIGAZ, already during the 11th Five-Year Plan, the capacity of electrically driven compressor stations will increase from 3 to 7.3 million kW, which will permit maintaining the relative contribution of electrical drive to the GPA stock in spite of the large numbers of GPA that will be introduced.

The increase in the capacity of electrically driven GPA will be ensured by the introduction of previously assimilated 4- and 12.5-MW plants, as well as plants with a capacity of 25 MW introduced during the 11th Five-Year Plan. The production of electrically driven 16-MW GPA is planned to begin during the 12th Five-Year Plan.

The period being examined will be a period of extensive introduction of GPA with full-head centrifugal superchargers. Up to 90% of the new compressor capacity will be based on full-head GPA. The introduction of such plants will increase the reliability and capacity of gas pipelines, decrease the requirements for GPA, simplify the technological design of compressor stations, and decrease energy consumption of compression.

Full-head superchargers are labelled according to the replaceable flow-through part by the degree of compression 1.35–1.37, 1.45, 1.52–1.55.

The large domestically produced refrigeration systems have electric drive. Their use is not always justified for the conditions of the gas industry. In this connection, the use of ATP-5-16/1 plants with electric drive and TKA-P-6,3/10 driven by a NK12-S T aircraft gas turbine is promising. The use of such plants will decrease the capital investment and reduce costs approximately by 20% compared with propane plants.

The presence of solid and liquid mechanical impurities in the transported gas decreases the capacity of the gas pipeline and causes significant erosive wear of the equipment in compressor stations.

Thus, for example, due to erosion of the supercharger, its polytropic efficiency decreases by 4–10%, which, in its turn, leads to the same overconsumption of fuel gas.

Installation of fully-equipped–block systems for preparing fuel, startup, and pulsed gases, dust catchers, and filter-separators will ensure a high degree of removal of mechanical impurities from the gas, will decrease the erosive wear of equipment, and will decrease the overconsumption of fuel gas.

Systems utilizing heat from exhaust gases are widely used in mainline compressor stations in the USSR. At the present time, the main method for utilizing the thermal secondary energy resources (SER) is for obtaining hot water, primarily for heating purposes.

For the future, up to the year 2000, a program is planned for more complete utilization of SER for creating hot-house-vegatable combines.

The production of a unified heat-recovery section for obtaining hot water, based on which recovery heat exchangers for GPA are produced, has been assimilated. It is expected that 3500–4000 heat exchangers for all types of domestic GPA will be produced over the planned period.

4. Ensuring the Reliability and Flexibility of Gas Transportation Systems

One of the most important aspects of planning of the development of UGSS up to the year 2000 with the present trend in its development is providing systems properties that meet the requirements of economic and operational efficiency as well as reliability (including viability) of gas supply. In addition, among the various measures ensuring the reliability and flexibility of UGSS operation, together with the traditional methods (increase in the reliability of pipes and equipment, backup of plants at compressor stations, construction of ties between strings, etc.), systems method for ensuring the network-structural reliability and interaction of different subsystems (main gas pipelines, underground gas storage (UGS)) are becoming increasingly urgent.

Assurance of reliability and flexibility of the gas transportation system is based on the following principles:

To ensure network-structural reliability, in planning the development of UGSS, the consolidated indicators should include the requirement of viability and flexibility of the system, i.e., its capability to withstand strong perturbations, for example, related with a sharp decrease in the capacity of separate large gas pipelines. On this basis, Tyumen gas is introduced along several routes [6]: central, western, and southern.

To ensure reliability of gas transportation and uninterrupted delivery to consumers, it is necessary to start with the complex nature of the solutions to technical problems, foreseen in the planning of separate gas pipelines. Each planned large main gas pipeline is no longer viewed as being autonomous, but rather as a stage in the development of the UGSS as a whole and its reliability is also ensured by including the interaction with other objects in the system. It is from this point of view that the sizes and distribution of gas reserves stored underground, points of connection of each gas pipeline with the UGSS for operational maneuvering of gas flows, production capacities of gas enterprises, as well as the sizes and distribution of production capacities of maintenance and repair systems are determined in providing backup mainline capacity.

An important method for increasing the reliability of UGSS is to increase the scientific-methodological level of analysis and development of solutions to technical problems ensuring gas transportation already at the planning stage of gas pipelines, since inclusion of reliability factors has an important effect on the optimal parameters of the planned object and all of its technical-economic indicators. For this purpose, in 1980, a new procedure for calculating gas pipeline reliability, developed by VNIIGAZ and the I. M. Gubkin Institute of the Petroleum and Gas Industry in Moscow and reflecting the present level of methodological and algorithmic development in this field, was developed and confirmed by the Ministry of the Gas Industry of the USSR [24].

In accordance with this procedure, the starting data for calculating the reliability of main gas pipelines are the indicators of reliability and maintenance service on the main technological equipment at compressor stations and pipelines in the linear part.

These indicators are taken from average statistical actual operational data, including the trend in the improvement of the quality of construction of equipment and system maintenance. In addition, all basic indicators are estimated in a differential manner for the different regions of the country. Since the specific conditions of Western Siberia have a special effect on the reliability indicators, this effect is taken into account by special correcting coefficients. The method relies on an appropriate complex of algorithms and programs.

It is this method that was used to calculate the reliability indicators for main gas pipelines to be put into operation during the 11th and 12th Five-Year Plans.

An example of the selection of an optimum technological scheme and gas pipeline parameters including solutions to scientific-technical problems that provide for system reliability is the Ureng–Pomary–Uzhgorod gas pipeline project, which enters into the general scheme of this gas pipeline were determined using a combination of different methods and means for providing backup, including also systems means and their effect on technological-economic indicators.

The problem of assuring the reliability of gas transportation and continuity of gas supply along the Ureng–Pomary–Uzhgorod gas pipeline was solved in the plan by studying its system reliability, when together with technical aspects, an increase in the reliability of the gas pipeline itself and its interaction with UGSS were included. Reliability of gas transportation is ensured by the following means:

backup of the main gas pipeline by choosing an efficient backup capacity of GPA at compressor stations, construction of ties between parallel pipeline strings, construction of ties between compressor plants in multiplant compressor stations with the same type of gas pumping aggregates for maneuvering backup GPA;

creation of operational gas reserves in UGS;

creation of points at which the gas pipeline is connected with UGSS for operational maneuvering of gas flows.

All solutions to technical problems foreseen in the plan of the Ureng–Pomary–Uzhgorod gas pipeline, concerning the assurance of gas transportation and continuity of gas supply for export, taken as a whole, greatly increase the reliability of the pipeline. This increase in reliability corresponds to an increase in the average annual capacity of the gas pipeline by approximately 4% compared to the gas pipeline variant that does not include means for increasing pipeline reliability.

The reliability of other main gas pipelines, which enter into the development of the UGSS up to the year 2000, is also included in an analogous manner.

An important means for increasing the reliability of UGSS is increasing the reliability of the elements of gas transportation systems (linear part and compressor stations along the pipelines).

An increase in the length, capacity, and working pressure in gas pipelines increases the requirements for reliability of compressor station equipment. Calculations show that decreasing the down time of gas pumping plants by 1% (of the calendar time) increases the annual productivity of the gas pipeline by 1.5%.

An important reserve which increases the reliability of GPA and compressor stations as a whole is improvement of the repair work. Taking into account the increasing complexity of construction of new GTI, the level of repair work must be increased by making it easier to repair structures, by developing methods and means for technical diagnostics of GPA, and performing the repair work on aggregated units at manufacturing plants.

Modular (unit) maintenance makes possible large savings in resources and decreases the need for spare parts and transportation expenses. In modular maintenance, defective modules are replaced in a standard manner independent of the operating time.

Defects in modules are revealed by technical diagnostics. Modules are repaired in a centralized manner at the GPA manufacturing plants. The possibility of modular maintenance presumes imposition of special requirements on the construction of GPA concerning the ease of monitoring and repair, as well as the need for special equipment for assembly and disassembly of modules at compressor stations. The transition to the new system for technical maintenance will begin during the 12th Five-Year Plan. Diagnostics of gas pumping plants at compressor stations should provide the ability to monitor the state of plants without shutting them down and disassembling them. Finally, it is becoming possible to change over from the system of planned-preventative maintenance, for which the periodicity and amount of repairs are established according to average group indicators, to a repair system that follows the actual state of specific GPA. The introduction of technical diagnostics at compressor stations decreases the operational expenses for compressor stations by 5–10%. The savings is achieved as a result of the following:

decrease in down time of GPA due to forecasting and prevention of failures;

elimination of unjustified periodic disassembly of plants;

decrease in labor-intensiveness of repair and technical maintenance of GPA;

increase in available capacity and efficiency of GPA;

decrease in the number of repair and service personnel;

improvement in planning the number and supply schedule of required spare parts.

The methods and means for technical diagnostics of practically all types of GPA will be introduced during the 12th Five-Year Plan.

5. Conclusions

The program for decreasing the consumption of metal in the construction of main gas pipelines presumes the construction of gas pipelines using pipes with the most economically efficient diameter (1420 mm), new economical brands of steel with improved strength characteristics for making pipes, as well as new types of pipes: multilayer, double-layer, spiral-seam, banded, and others.

References

1. Khodanovich, I. E., Z. T. Galiullin, and B. L. Krivoshein (1973). "Investigation of technological regimes of main gas pipelines," *Geologiya, razrabotka, transport, khranenie i pererabotka priridnogo gaza (Geology, Development of Fields, Transportation, Storage, and Refining of Natural Gas)*, VNIIGAZ, Moscow, pp. 93–101.
2. Vasil'ev, Yu. N., Z. T. Galiullin, V. S. Zolotarevskiĭ, *et al.* (1973). "Installation for cooling mainline gas," Inventor's Certificate 383974 (USSR); *Byull. Izobr.*, No. 4.
3. Galiullin, Z. T., G. É. Odishariya, and N. I. Ozotov (1980). *Kholodil'nye ustanovki v sistemakh magistral'nogo transporta gaza (Refrigeration Installations in Mainline Gas Transportation Systems)*, VNIIÉgazprom, Moscow.
4. Sinitsyn, S. N., and E. V. Leont'ev (1974) "Effect of nonuniformity in gas inflow on the technological scheme of a compressor station with centrifugal superchargers," *Trudy VNIIGAZa, Razrabotka gazovykh mestorozhdeniĭ, transporta gaza (Proceedings of VNIIGAZa, Development of Gas Fields and Gas Transportation)*, Moscow.
5. Sinitsyn, S. N., and I. V. Bartsev (1974). "Optimal range of nominal parameters of compressor stations on mainline gas pipelines." *ibid.*
6. Orudzhev, S. A. (1981). *Goluboe zoloto Zapadnoĭ Sibiri (Blue Gold of Western Siberia)*, Nedra, Moscow.
7. Galiullin, Z. T., E. V. Leont'ev, and S. Kh. Neĭtur (1982). "Effect of design parameters of gas pipelines on the energy intensiveness of gas transportation," *Gazovaya promyshlennost'*, No. 3.
8. Odishariya, G. É. (1976). "Effect of the depth of cooling of the gas on the technical-economic indicators of gas transmission," *Gazovaya Promyshlennost'*, No. 8, 45–48.

9. Gudkov, S. F., O. A. Ben'yaminovich, and G. É. Odishariya (1970). "Efficient regions of application of different technological schemes for transportation of large volumes of natural gas," *Report at the 11th International Gas Congress*, Moscow.
10. Galiullin, Z. T. (1978). "Methods for increasing the operational reliability of gas transportation systems," *Gazovaya Promyshlennost'*, No. 10, 30–34.
11. Galiullin, Z. T. (1979). "Development and assurance of the reliability and efficiency of transportation of large gas flows along main gas pipelines", *Nauchno-teknicheskiĭ progress v dobyche i transporte gaza (Scientific-Technical Progress in the Production and Transportation of Gas)*, Nedra, Moscow, pp. 93–102.
12. Ivantsov, O. M. (1979). "Problems of main gas pipelines in the North," *Izv. Akad. Nauk SSSR, Energ. Trans.*, No. 4, 3–14.
13. Khodanovich, I. E. (1961). *Analiticheskie osnovy proektirovaniya i ékspluatatsii magistral'nykh gazoprovodov (Analytic Foundations of the Design and Operation of Main Gas Pipelines)*, Gostoptekhizdat.
14. Anderson, I. H. (1965). "Liquefied methane pipeline: next gas transmission," *Ster. Oil and Gas J.* **63**, No. 6, 67.
15. *4th International Conf. on LNG, June 24–27*, 1974, Session 7.
16. Belousov, V. D., Z. T. Galiullin, and V. I. Chernikin (1961). "Optimum parameters of multistring main gas pipelines," *Gazovaya Promyshlennost'*, No. 3.
17. Galiullin, Z. T., and V. I. Chernikin (1964). *Novye metody proektirovaniya gazonefteprovodov (New Methods for Designing Gas and Oil Pipelines)*, Nedra, Moscow.
18. Karpov, S. V., N. A. Karpova, and Z. T. Galiullin *et al.* (1978). *Osnovnye puti razvitiya magistral'nogo transporta gaza (Main Paths in the Development of Mainline Gas Transportation)*, VNIIÉgazprom, Moscow.
19. Galiullin, Z. T., A. I. Garlyauskas, I. E. Khodanovich, and É. S. Salimzhanov (1970). "Principals of optimization of complex gas-supply systems," *Trudy VNIIgaza, Voprosy transporta prirodnogo gaza (Proceedings of VNIIgaza, Problems in Transportation of Natural Gas)*, Nedra, Moscow, Nos. 38/46.
20. Sinitsyn, S. N., and E. V. Leont'ev (1970). "Search for optimum of the gas-pipeline–compressor-station system," *ibid.*
21. Sukharev, M. G., and E. R. Stavrovskiĭ (1975). *Optimizatsiya sistem transporta gaza (Optimization of Gas Transportation Systems)*, Nedra, Moscow.
22. Dobrokhotov, V. D., A. K. Klubnishkin, T. T. Pyatakhina, and A. N. Kaluzhskikh (1970). "Input control devices for natural gas centrifugal superchargers," *Refer. inform., ser. Transport o khranenie gaza (Review Series on Gas Transportation and Storage)*, VNIIÉgazprom, Moscow, No. 6.
23. Sinitsyn, S. N., and E. V. Leont'ev (1980). "Expedience of staged introduction of capacity of compressor stations under construction," *ibid.*, No. 3.
24. *Metodika rascheta nadezhnosti magistral'nykh gazoprovodov (Procedure for Calculating the Reliability of Main Gas Pipelines)*, VNIIgaz, Moscow (1980).

Sov. Tech. Rev. A Energy Reviews, Vol. 2, 1985, pp. 111–154
0275–7893/85/002–111 $30.00/0

THE UNIFIED GAS SUPPLY SYSTEM OF THE USSR

A. D. SEDYKH and A. I. GRITSENKO

Ministry of the Gas Industry of the USSR, 117939, Moscow, V-31, Ulitsa Stroitelei, 8, Korpus 1, USSR; and All-Union Scientific Research Institute of Natural Gases, 142700, Moscow Province, Gor. Vidnoe, GSP, USSR

Abstract

The gas industry is one of the most important sectors of the fuel-energy complex of the USSR. It is based on a unified system for supplying gas that consists of all technologically interrelated objects for production, mainline transportation, storage, refining, and distribution of gas.

In this paper, the most important stages in the formation and development of the unified gas supply system and its separate elements are examined. The basic provisions of the technical policy of the industry concerning the problems of improving the technology and the technical means for developing gas and gas-condensate fields, and the treatment, transportation, underground storage, and refining of gas are formulated.

Contents

1. Introduction

The gas industry is a comparatively young sector of the Soviet economy. Mainline gas transportation began in the Soviet Union during the Second

Table I Production of the Basic Forms of Energy Raw Materials and Generation of Electricity in the USSR for the Period 1965–1980

Form of fuel-energy resource	Years			
	1965	1970	1975	1980
Gas, billion cubic meters	128	198	289	435
Petroleum, million tons	248	353	491	603
Coal, million tons	578	624	701	716
Electricity, billion kW-h	507	741	1039	1295
Total production of energy resources, billion tons of reference fuel*	0.97	1.22	1.6	1.97
Relative fraction of gas, %	15.5	19.1	21.8	27.0

*The calorific value of the fuel equals 7000 kcal kg^{-1}.

World War with the construction of a gas pipeline with a diameter of 300 mm and a length of about 160 km in the region of Kuibyshev on the Volga. At the end of the war, construction began on another main gas pipeline from Saratov to Moscow, connecting the Povolozh'e gas fields with Moscow. This pipeline was constructed from 235-mm pipe, was 843 kilometers long, and transported 0.5 billion cubic meters of gas per year with the help of six compressor stations.

In recent years, the Soviet gas industry has developed at an accelerated rate, ensuring continuous growth of the relative contribution of natural gas to the energy balance of the country.

Data on the increase in the production of gas and other fuels in the USSR for 1965–1980 are presented in Table I. It is evident that during this period the production of gas, oil, and coal increased by factors of 3.4, 2.48, and 1.2, respectively, and the generation of electricity increased by a factor of 2.55. Thus, from 1965 to 1980, gas production grew 1.33 times more rapidly than the generation of electrical energy, 1.37 times more rapidly than oil production, and 2.83 times more rapidly than coal production.

Natural gas is widely used as a universal fuel and a valuable chemical raw material in industrial production and it is an important factor in technical progress and in the increase in the productivity of labor.

In 1981, in the USSR, about 93% of all cast iron, the same amount of open-hearth steel, about 50% of the rolled metal, 60% of the cement, more than 80% of the ammonia, 72% of the methanol, and almost 100% of the refractory materials were produced with the use of natural gas.

The level of domestic use of gas is very high. Eighty-two percent of cities and urban areas and 77% of rural areas use gas. About 200 million people in the Soviet Union use natural and liquified petroleum gas at the lowest price in the world.

The development of the gas industry in the Soviet Union can be separated into two characteristic stages [1].

The first stage encompasses the period from 1941 to 1955. This stage is characterized by the creation of the necessary prerequisites for the gas industry to become an independent sector of the economy.

The second stage in the development of the industry (1956–1972) is characterized by such a large increase in the amount of construction for producing, transporting, and refining gas and condensate that it was necessary to create a separate ministry for construction in the oil and gas industry. This resolution was adopted by the Central Committee of Ministers of the USSR in 1972.

At the present time, the gas industry in the USSR is at the third stage of its development, The distinguishing features of this are as follows:

large source of raw material:

advanced technology and techniques for producing and treating gas, pipeline transportation, storage, and refining of gas and gas condensate;

large, organized industrial-scale production of materials and equipment for gas fields, main pipelines, gas refineries, and other industry objects;

highly qualified engineer-technicians and scientists;

enormous economic-production potential, which permits independent and intensive development of the industry into the long-term future.

The planned program for the development of the fuel-energy complex provides for further increases in the relative contribution of natural gas under conditions of increasing production of energy raw materials and generation of electricity in the country. In accordance with the resolutions adopted at the 26th Congress of the Communist Party of the Soviet Union [2], gas production in 1985 will be increased to 600–640 billion cubic meters with an average annual increase in production equal to 35–40 billion cubic meters. At the same time, more than 75% of the total volume of gas produced will come from the gas and gas-condensate fields in Western Siberia.

To provide an uninterrupted supply of gas for the national economy, a unified gas supply system (UGSS), which is the largest and most efficient type of system in the world, has been created in the USSR. This system includes 250 gas and gas-condensate fields, an interconnected network of 140 thousand kilometers of main gas pipelines, more than 300 compressor stations in which more than 3.3 thousand gas pumping aggregates with a total power of more than 20 million kilowatts has been installed, underground storage with a large volume of reactive gas, and an enormous number of stations for distributing gas in more thant 156 thousand cities and population centers.

The successful operation of the UGSS in the USSR is made possible by nological system, which must meet the requirements of the national economy for gas both as a fuel and as a raw material for the chemical industry at the lowest possible cost. This system has evolved according to its own rules and peculiarities.

The successful operation of the UGSS in the USSR is made possible by the socialist economic system. Public ownership of the natural resources of the country and of the means of production has provided for optimum and rapid development of the UGSS.

The unified gas supply system is evolving in close connection with the development of the national economy as a whole. The development of the national economy determines the rate of growth in gas production and the direction of gas flow. Scientific-technical progress in machine building, metallurgy, instrument building, and other sectors of the economy determines the constantly increasing technical level of the gas supply system. In its turn, the UGSS has a decisive effect on the development of the economy. The use of natural gas in the leading sectors of the economy has a direct effect on technical progress and increases the productivity of labor. The development of large gas transportation systems greatly affects the distribution of energy-intensive industries. Increased utilization of gas also leads to appreciable social conseqeunces: the productivity of labor increases, the number of people working in other less efficient sectors of the fuel-energy complex decreases, and environmental damage also decreases.

The characteristics of the main elements of the unified gas supply system, including the supply of raw material, are presented below.

2. Raw Material Supply

The foundation for the development of the unified gas supply system is a reliable supply of raw material, discovered by Soviet geologists, for the gas industry. The successes achieved here are due to the extensive development of geological-exploratory work in different regions of the country and especially in Western Siberia and Central Asia during the last ten years.

During the early years of the Soviet government, the main exploratory work was conducted primarily in the petroleum regions of Azerbaĭdzhan SSR, Grozenensk and Ural-Embensk provinces, and Western Turkmen, as well as in the southern regions of Krasnodar Kraĭ and Dagestan, This led to the discovery of fields with relatively small gas and condensate reserves. In addition, based on these data, it was incorrectly concluded that the resources of natural gas in our country were severely limited.

During the Second World War, geologists explored the oil and gas resources in the regions of Povolzh'e and Komi ASSR. The first gas fields, which served as the foundation for the development of the gas industry during the first ten years following the war, were opened up here. Large gas and gas-condensate fields were later opened up in the Ukraine, Northern Caucasus, Uzbekistan, and other regions of the European part of the USSR and Central Asia [3].

Commercial flow of gas in 1963 from the exploratory well in the settlement of Berezovo in the province of Tyumen, which marked the opening of the Western Siberian oil and gas bearing province, was a great achievement of the exploratory work.

These discoveries proved that gas fields exist in the most diverse regions of our country: in the European part of the USSR, in Western and Eastern Siberia, and in Central Asia. The inclusion of new territories into the search and the large increase in the volume of geological exploratory work led to the discovery of new gas-bearing provinces, which rapidly increased gas reserves in the USSR.

Unique natural gas fields have been found in the norther part of Tyumen province: Ureng, Yamburg, Zapolyarnoe, Medvezh'e, and others. There was a large increase in reserves in Central Asia, mainly due to the development of the Shatlyk, Naip, Achak, and Urtabulak fields. The Vuktyl' and Orenburg gas-condensate fields, a group of gas-condensate fields in the Krasnodarsk Kraĭ, as well as the Efremov and Krestishchen fields in the Ukraine were opened up in the European part of the country. During the period from 1956 to 1972, the proven reserves of natural gas increased by more than a factor of 25. Gas production in the country correspondingly increased from 12 to 221.4 billion cubic meters. In 1956, practically all gas was produced in the European part of the USSR, but in 1972, the relative contribution of this region to total production in the USSR decreased to 66%.

Gas reserves continued to increase in the 1970s. The large gas-bearing potential of the Yamal Peninsula in northern Western Siberia, the northern regions of Timano–Pechor province, and the western regions of Uzbekistan and other territories was confirmed. The Layavozh and Vaneĭvis fields in the province of Arkhangel'sk, the Bovanenkovskoe and Yuzhno–Russkiĭ fields in Tyumen province, the Shurtan field in Usbekistan, and Sobolokh–Nedzhelin field in Yakutsk ASSR, and other large fields were discovered during this period.

A large number of gas, gas-condensate, and gas-oil fields, situated in different regions of the country and confined to deposits from the Cambrian to the Neogene periods inclusively, have been discovered in the USSR. The largest accumulations of gas, comprising 86% of proven reserves, are concentrated in Mesozoic deposits; 12% and 2%, respectively, are concentrated in Paleozoic and Cenozoic deposits.

In the European part of the USSR, proven gas reserves are concentrated in the Komi ASSR, and the province of Arkhangel'sk, in Ural-Povolzh'e, in the Ukrainian SSR, and in the Northern Caucasus. The Vuktyl (Komi ASSR), Zapadno-Krestishchen (Ukrainian SSR), Astrkhan (Astrakhan province), and Orenburg (Orenburg province) fields are the largest fields.

The Orenburg gas-condensate field was opened up in 1956 and development of this field began in 1971.

The gas in the Orenburg field contains a considerable amount of ethane, propane, butanes, hydrogen sulfide, and condensate. The hydrogen sulfide content of the gas varies from 1.4 to 3.9%. The useful components of the gas are recovered at the Orenburg gas-chemical complex.

The gas reserves in the Republics of Central Asia are concentrated mainly in Eastern Turkmen and Western Uzbekistan. The Achak, Naip, Shatly, Kirpichlin, and Sovet-abad in Eastern Turkmen and Shurtan, Zevardin, Gazlin, and other fields in Western Uzbekistan are the largest fields.

Gas from the Central Asian Republics is delivered to the European regions of the country and the Ural regions, as well as to Kirgiziya and Kazakhstan. In recent years, the large Sovet-abad gas-condensate field was discovered and explored in this region. The gas in this field is about 95% methane and 1.3–1.5% heavy homologs. The condensate content of the gas is 5.8–15.9 $g\,m^{-3}$.

The Western Siberian oil-gas-bearing province occupies a territory with a total area of about 2.5 million square kilometers, of which 1.8 million square kilometers have potentially promising oil and gas deposits [4].

Most of the gas reservoirs in this region are confined to three oil-gas bearing formations: Aptian–Cenomanian, Neocomian, and Jurassic.

The Aptian–Cenomanian formation consists of frequent alternation of weakly cemented sandstances and aleurolites of continental and offshore-marine origin.

The Cenomanian deposits are associated with the gas reservoirs in the Ureng, Zapolyar, Yamburg, and Medvezh'e fields.

The Neocomian formation represents alternating layers of sandstones, aleurolites, and clays characterizing a sharp lithological variability across the geological cross-section.

The highest concentrations of gas reserves in northern Tyumen province occur in the Ureng and Yamal gas-bearing regions. The Ureng region includes the Nadym-Pur and Pur-Tazov interfluve.

Several gas fields have now been discovered in Eastern Siberia and the Far East. This extensive territory has not yet been adequately studied. New, large gas fields are expected to exist there.

The natural gas from many of the fields contains, in addition to methane, other valuable components such as ethane, propane, butane, sulfur, and condensate.

Gas reserves with a high ethane content (more than 3% by volume), useful for chemical reprocessing, have been extensively developed in the European part of the country and are also characteristic of the Lower Cretaceous and Jurassic deposits of Western Siberia, Eastern Turkmen, and Western Uzbekistan and the Paleozoic deposits of Eastern Siberia. The hydrogen-sulfide-containing gases are concentrated mainly in the regions of Ural-Popolzh'e and Timan-Pechor and in Central Asia.

Intensive field development and the relatively small increment to the gas reserves in the European part of the country has led, in recent years, to a considerable decrease in the proven reserves in a number of regions. As a result, the geographic distribution of gas resources has changed considerably. During the mid-1960s, the main proven reserves of natural gas were concentrated in the European part of the USSR (Northern Caucasus, Ukraine, Povolzh'e). Now most of the total reserves of the USSR are concentrated in the Asiatic region of the country and, primarily, in northern Tyumen province.

The predominant contribution of Tyumen gas in the raw material supply of the gas industry has determined the east-to-west flow of gas in the national economy. In its turn, the gas supply system that has evolved greatly affects the development of the raw material supply of the country. In particular, efficient use of the gas transport system requires further geological-exploratory work in order to open up gas fields situated near existing gas pipelines,

Thus the intimate relation between the gas supply system and the raw material supply is one of the characteristics of the evolution of the UGSS.

3. Gas Production

The production of natural gas within the territory of the USSR began in the 1920s in the western Ukraine in small quantities. The development of the comparatively small fields of Komi ASSR and Povolzh'e began in the

Table II Principal Indicators of Natural Gas Production in the USSR

Indicators	Years									
	1965	1970	1975	1976	1977	1978	1979	1980	1981	1985
Gas production, billion cubic meters	127.7	197.9	289.3	321.0	346.0	372.2	406.6	435.2	465.3	630
including: natural gas	111.2	175.0	260.7	290.0	314.5	338.3	373.0	402.1	430.3	590
petroleum (casing-head) gas	16.5	22.9	28.6	31.0	31.5	33.9	33.6	33.1	35.0	40.0
Operational stock of gas wells (at the end of the year), wells	2385	3964	5532	5780	6032	6352	6662	6981	7253	–
Average daily production rate of gas wells, thousand cubic meters	180	149	151	159	166	173	180	185	224	–
New wells put into production, wells	335	456	447	434	434	479	456	441	446	–

1940s. Gas production began to develop especially intensively in the 1950s, when assimilation of the Shebelin, Severo-Stavropol'skoe, and Gazli Fields started. In 1960, gas production (natural and casing-head gas) already constituted 45.3 billion cubic meters [1]. The main gas production indicators since the 1960s and up to the present time are presented in Table II.

At the present time, the development of the gas industry is characterized by high rates of growth of gas production. Thus, in 1971–1975, the average annual increment constituted 18.3 billion cubic meters, but in 1976–1980, the increment increased by a factor of 1.6 and reached 29.2 billion cubic meters. In 1980, 435.2 billion cubic meters were produced and 465.3 billion cubic meters were produced in 1981.

The Republics of Central Asia are some of the largest gas producing regions of the country. The existing raw material supply of this region not only permits maintaining the level of production reached, but it also provides for further growth in production in the immediate future.

The level of gas production in Uzbekistan is maintained due to the development of fields that have already been discovered in the western parts of this region. The difficulties in assimilating this region are due to the difficult geographical (desert nature of the location, inadequate transportation network) and mining-geological (presence of corrosive components in the gas from the main fields, anomalous temperature-and-pressure conditions in separate reservoirs, gas shows, etc.) conditions.

An important gas-producing region is the Orenburg gas-condensate field with the very large gas-chemical complex, where gas is produced and refined, yielding a high-purity elemental sulfur, the propane-butane fraction, a stable condensate, and other very valuable products required by the chemical and petrochemical industries.

Experience in operating this complex will play an important role in assimilating the Astrakhan field during the period 1981–1985. The gas from this field contains up to 25% hydrogen sulfide by volume.

As before, the old gas producing regions, such as Azerbaĭdzhan SSR, Komi ASSR, and others, will play a large role in gas production. Gas production will be maintained here by increasing the gas recovery factor via an increase in the capacities of the booster compressor stations and extensive use of means for prolonging the lifetime of flooded wells.

The present stage of development of the gas industry of the USSR is characterized by the displacement of the center of gas production into Western Siberia, which is a remote region with difficult access and very difficult environmental and climatic conditions. The Western Siberian gas production region is distinguished by rapid and large growth. In assimilating the largest fields (Medvezh'e and Ureng fields), the complex of fundamentally new scientific-technical problems involved in establishing production and transportation of gas under swampy and permafrost conditions was solved for the first time in the USSR and in world practice [5].

By 1985, gas production in this region should reach 330 billion cubic meters. An important characteristic of this region is the high concentration of gas reserves. Of the production volume planned for Western Siberia, the largest levels concern the Ureng field. In the immediate future, Western Siberia will also be a large supplier of gas condensate. Large quantities of ethane and liquified gases (propane–butane mixture) will also be produced here. An experimental commercial extraction of oil from the margins of the gas-condensate reservoirs is planned.

The basic problems facing scientific-technical progress in the production of gas at the present time originate with the following characteristics of the development of this industry:

presence of appreciable gas reserves in a small number of fields (concentration of reserves);

location of most large gas fields that have been opened up in regions with poor accessibility and difficult environmental and climatic conditions;

presence of impurities, such as gaseous condensate, helium, hydrogen sulfide, and carbonic acid in the gas from a number of fields;

transition to the final stage of development of most fields in the European part of the country;

assimilation of gas reservoirs and fields with anomalously high formation pressures.

These characteristics make it necessary to conduct investigations of the development of fields in several directions.

The first direction is to increase the degree of extraction of valuable components from the mineral resources.

Here, it is primarily necessary to concentrate the production capacities with special well-placement systems, high well productivity, and unified production lines in setups for over-all preparation of gas. Such systems are now in place in fields in northern Tyumen province: Medvezh'e, Ureng, and other fields.

One of the most complex probelms is to increase the condensate and oil yield in developing gas and oil-and-gas-condensate deposits. Studies have affirmed the possibility of extracting precipitated condensate and oil from the periphery with the help of different hydrocarbon solvents.

The second direction involves the creation of reliable systems for controlling gas-production processes. Previously, in developing comparatively small fields in "old" gas-bearing regions (Stravropol, Krasnodarsk Kraĭ, Komi ASSR, Ukraine, Uzbekistan, and others), these fields could be easily controlled with appropriate well placement and operating regimes. In exploiting such fields as Orenburg, Medvezh'e, Ureng, Yamburg, and others, these two measures are already clearly inadequate. An example of the solution of control problems is one of the promising methods for monitoring the movement of the gas–water interface. To perform such monitoring under the conditions in the Ureng field using only observational wells, a large number of such wells is required. In this connection, there arises the problem of creating promising economically-practical in-well methods, one of which is performing regular high-flow gravimetric investigations of mass transfer processes in the formation.

The third direction for improving gas-production technology is to create efficient methods for designing gas production objects, for example, an object such as the "formation–surface-construction–mainline" system or even a more complicated object, when such a system includes gas refineries.

The following basic variants of man-machine systems must be developed for long-range planning and scheduling of gas production und.r conditional uncertainties in available information:

for large gas-bearing regions, taking into account the basic objects in the production and social infrastructures;

for separate base fields, which determine the load on the main gas pipelines in the region;

for groups of fields, which comprise the raw material supply of the gas-chemical complexes [6].

A broad group of scientists and specialists from a number of ministries and departments, with whose combined efforts new solutions to scientific and technical problems are being introduced, thereby ensuring that the planned rates of growth with maximum reduction of capital investments are realized, has been directed to realize this very important economic program.

By now, reliable methods for producing and preparing gas for transportation as well as an industry standard for natural gas entering the mainline, in which, based on optimum technical-economic indicators, limits have been imposed on the water and hydrocarbon dew points and on the hydrogen sulfide content of the transported gas, have been developed and put into practice. The introduction of new solutions to technical problems involved in the production and preparation of gas for transportation and the creation of a standard for natural gas have resulted in highly efficient operation of main gas pipelines.

To ensure high rates of increase of gas production, fully-equiped block assembly of technical equipment has been developed and put into use. This has made it possible to change over to industrial methods for constructing installations for complex preparation of gas for transport (ICPG) and thereby to decrease the time required to build the installations by a factor of 2–3. The standardized series of ICPG production lines that has been developed (Table III) encompasses practically all variants that could be encountered in equipping fields. The use of standardized installations makes it possible to achieve large savings as a result of optimizing the solutions to design problems.

To decrease expenditures on equipping fields and to decrease disruptions in putting these fields into operation, a great deal of attention is devoted to concentration of production. The new system for placing wells in the field and the centralized system for collecting gas, built as a linear path with group tie-in of wells and developed for this purpose, has made possible a reduction in the amounts of metal used in the delivery lines, creation of ICPG with a production capacity of 15–20 billion cubic meters per year (as opposed to 3–7 billion cubic meters per year), as well as a 3–5-fold reduction of the total volume of construction and assembly work (depending on the number of wells connected).

An important problem arising in gas production is protection of pipes and gas-field equipment from corrosion caused by the presence of corrosive components, such as carbon dioxide, hydrogen sulfide, hard formation water, and organic acids present in the gas produced, as well as from soil corrosion.

As the well depth increases, the number of gas and gas-condensate fields with corrosive media increases and the degree of corrosiveness also increases due to an increase in pressure, temperature, and content of corrosive components in the extracted product.

In this connection, at the end of the 1960s and beginning of the 1970s, the gas industry in the USSR was faced with a new problem of ensuring the operational reliability of equipment and pipes used in production, preparation, transportation, and refining of gas containing hydrogen sulfide.

Pipes and equipment resistant to hydrogen sulfide were not produced on commercial scales in the USSR up to the end of the 1960s.

Based on the results of studies, seamless gas pipes, apparatus, fixtures, and other equipment produced domestically from carbon steels containing

Table III Technical Specifications of Plants for Complex Preparation of Gas for Transportation

Nomenclature	Capacity, million normal $m^3 s^{-1}$	Working pressure, MPa	Specific input of materials, kg/thousand normal $m^3 day^{-1}$	Occupied area (according to dimensions of blocks), m^2
Production; line for low-temperature separation of gas	1	8–10	102	113
	3	8–10	52	133
	5	8–10	61	138
Production line for absorption drying of gas	3	8	34	76
	5	8	13	41
	10	8	9	41

up to 0.2% carbon (Nos. 20 and 10 steel) and from corrosion-resistant steels alloyed with chrome, nickel, molybdenum, and titanium were used. In addition, to avoid cracking caused by hydrogen sulfide, working stresses were limited by greater wall thicknesses and inhibitors were used for additional protection against general corrosion.

When possible, the gas was dried, eliminating the formation of films on pipe and equipment walls, which eliminated electrochemical corrosion of the interior surfaces.

Experience of more than eight years in exploiting gas fields containing hydrogen sulfide has demonstrated the satisfactory resistance of most pipes and equipment used to cracking caused by hydrogen sulfide.

Based on scientific research, a system of measures has now been developed for protecting pipes and production equipment and equipment for field preparation, transportation and processing of gas containing hydrogen sulfide.

This system includes the use of low-alloy and carbon steels in manufacturing the pumping-compressor, casing, and gas-line pipes, the apparatus for field preparation and refining of gas, and the structural parts of the flow and shut-off control fittings so as to make them resistant to hydrogen-sulfide cracking, as well as the use of corrosion inhibitors and gas drying.

High-alloy steels are used to make seals for the shut-off valve fittings, tube bundles, and apparatus operating at high temperatures in refining gas and gas-condensate containing hydrogen sulfide, instruments, etc.

In the future, the system and the means for protection indicated above will have to be improved by increasing the reliability and economic practicability, taking into account the opening of the new gas fields with a higher content of carbon dioxide and hydrogen sulfide, large well depths, and high formation pressures and gas temperatures.

An important protective measure is efficient inhibition, which is one of the simplest and most effective methods for fighting against corrosion. Corrosion inhibitors slow down and, in many cases, completely stop the destruction of metals in corrosive media. The advantage of this method is the possibility of using it without introducing significant changes into the technological processes in the production and transportion of gas. The extent to which inhibition of gas wells has been introduced in the USSR can be judged from the following data: in 1965, only five wells were inhibited; in 1968, 173 wells were inhibited; and, in 1971, 250 wells were inhibited. In 1980, corrosion inhibitors were used in many hundreds of gas wells in different fields in our country: in Uzbekistan, Turkmen, Orenburg, and other regions. The use of inhibitors in these fields has reduced the rate of corrosion, which has greatly increased the operating time of wells between maintenance operations with insignificant expenditures on inhibition.

At the present, time corrosion inhibitors are used to protect gas wells, equipment in installations for preparation of gas and utilization of surface waters, gas pipelines for unscrubbed gas, and installations for removing sulfur.

Inhibition of wells in gas fields is accomplished by introducing the inhibitor into the space outside the pipes (continuously or periodically), as well as by pumping the inhibitor into the formation. In separate cases, an entire complex of measures is used including both periodic pumping of small quantities of the inhibitor into the formation and continuous introduction of the inhibitor at the bottom of the well.

At the present time, a series of highly effective corrosion inhibitors has been developed in the USSR, which aside from high protective properties, have good technological characteristics. Thus, for example, the IFKh Angaz-1 hydrogen-sulfide-corrosion inhibitor is also used as an antifoaming agent in gas scrubbing setups at refineries.

Future work includes developing new complex corrosion inhibitors for simultaneously preventing salt deposition, corrosion, and paraffin formation, which will not only decrease corrosion but will also increase gas production.

4. Mainline Gas Transportation

Main gas pipelines are the most important element of the unified gas supply system of the country. Mainlines determine the rates of development and the technical-economic indicators of the industry as a whole. More than 80% of the capital investment in the development of the gas industry in the USSR goes into the construction of main gas pipelines [7].

4.1. Basic Stages in Development of the System of Main Gas Pipelines in the USSR

The development of mainline gas transportation as a subsector of the gas industry began after World War II with the construction of autonomous gas pipelines from gas fields to the regions where the gas was used. The Saratov–Moscow gas pipeline was put into operation in 1946; the Dashava–Kiev 529-mm gas pipeline with a length of 202 km was put into operation in 1948 [8].

In the 1950s, main gas pipelines were introduced for transporting large quantities of natural gas from the fields in Stavropol and Krasnodar Kraĭs, and from the Saratov and Volgograd regions. Local gas pipelines were constructed in the Western Ukraine, in Kasnodar and Stravropol Kraĭs, and in the Volgograd and Saratov regions. In all, during these years, more than 6.5 thousand kilometers of main gas pipeline were laid. However, in spite of the considerable increase in gas pipeline construction, it was not necessary to construct interconnected gas pipelines at that time.

The role of pipeline gas transportation increased considerably in the first half of the 1960s, in connection with the vigorous development of the gas industry.

During this period, the large Northern Caucasus–Center gas transportation system and one of the largest, up to that time, gas mainlines were constructed: the two-string Bukhar–Ural gas pipeline with pipe diameters of 1020 mm pressurized at 5.5 MPa and a total length of 5000 km (including branches). A total of more than 30,000 kilometers of gas pipeline were laid during this period [9].

During the second half of the 1960s, the very large Central Asia–Center gas transportation system was put into operation. The first string in this system was constructed with 1029-mm gas pipe. During the construction of the second string, 1220-mm pipes were used for the first time in world practice.

At the end of the 1960s, intense assimilation of the large Vuktyl gas-condensate field began. The Vuktyl–Ukhta–Torzhok main gas pipeline was extended under difficult conditions over a short period of time. To transport gas from the Vuktyl field, a gas pipeline with a length of almost 1500 km with pipe diameter of 1220 mm was constructed. The Messoyakh–Noril'sk gas pipeline in Western Siberia and the Taas–Tumus–Yakutsk gas pipeline in Eastern Siberia were the first pipelines to be constructed under permafrost conditions.

During the same period, gas pipelines were constructed from Afghanistan and Iran. In all, during the second half of the 1960s, main gas pipelines were extended by approximately 24,000 km.

The development of the gas industry in 1971–1975 is characterized by the assimulation and development of large gas fields situated in northern Tyumen province, Central Asia, and Kazakhstan. During this period, the gas transportation system was enlarged by increasing the capacities of existing gas transportation systems. Two strings of the Medvezh'e–Nadym gas pipeline were put into operation during this period. Pipes with a diameter of 1420 mm pressurized at 7.5 MPa were used here for the first time in world practice.

The Medvezh'e–Nadym–Pung gas pipeline together with the operating Igran–Serov–Nizhniĭ–Tagil and the lower Tura–Perm–Gorki gas pipelines (with a length of more that 1700 km and a diameter of 1220 mm) formed the first course of the system from northern Tyumen province to the Ural–Center region. Gas began to flow to consumers through this system in 1972.

The third and fourth strings of the gas pipeline, with diameters of 1220 mm and 1420 mm, were put into operation in the gas transportation system from Central Asia to Center. During this period, the first string of the Orenburg–Aleksandrov Gaĭ–Novopskov 1220-mm gas pipeline, the Valdaĭ–Pskov–Riga 1020-mm pipeline, and the Torzhok–Minsk–Ivatsevichi 1220-mm pipeline were constructed. A total of more than 30,000 km of main gas pipelines was put into operation during the period from 1971 to 1975.

Multistring gas pipeline systems appeared at this stage of gas mainline construction in connection with the increase in gas flows. Combined operation of parallel gas pipelines connected by ties has increased the production capacity of the system by 4–5% as compared to independent operation of the pipelines. In addition, the ties between the strings increase the reliability of gas supply and decrease the short-fall of gas to consumers as a result of emergency or planned preventative maintenance work.

During the second half of the 1970s, the international Soyuz gas pipeline, extending from Orenburg to the western boundary of the USSR and constructed together with the member countries of the Council for Mutual Economic Assistance, was constructed and put into operation. This is the first gas pipeline of this length (2677 km) constructed in this country or abroad using 1420-mm pipe and pressurized at 7.5 MPa [10].

Another event, which was important for the gas industry in the USSR, occurred during this period. This was the introduction of gas from the Ureng field into the gas flow to the national economy. Gas from this field flowed mainly along the Ureng–Nadym–Ukhta–Torzhok–Ivatsevichi system with 1420- and 1220-mm pipe and the Ureng–Novopskov system, which consists of the following gas pipelines: the Komsomol'sk–Surgut–Chelyabinsk (line 1), Ureng–Chelyabinsk (line 2), and Chelyabinsk–Petrovsk 1420-mm lines and the Petrovsk–Novopskov 1220-mm line. The total length of the system exceeds 3500 km.

The third course of the Ukht–Torzhok gas pipeline, consisting of 1420-mm pipes, started operating during this period. With the introduction of this gas pipeline, gas from the Ureng field was given an outlet to western regions of the USSR and the member countries of Comecon.

A new Ureng–Pomar–Uzhgorod 1420-mm pipeline, pressurized at

7.5 MPa and with a length of 4451 km, for exporting gas is under construction. This pipeline is being constructed as part of an international project called "Gaz–Truby" [11]. The pipeline will be constructed and gas will be delivered in accordance with the time table established by the 'Gaz–Truby" contract [12].

The growth dynamics of the principal indicators of the main gas pipelines over the period 1960–1980 are displayed in Tables IV and V.

"The main trends in the economic and social development of the USSR from 1981 to 1985 and for the period up to 1990" [2] depend on the construction of large main gas pipelines with a high degree of automation and operational reliability. The following six large gas pipelines will be put into operation during the 11th Five Year Plan:

Ureng–Ukhta–Gryazovets, 1,440 km long with 16 compressor stations (CS);

Ureng–Punga–Petrovsk, 3,019 km, 24 CS;
Ureng–Punga–Nobopskov, 3,570 km, 30 CS;
Ureng–Pomary–Center (string 1), 3,423 km, 30 CS;
Ureng–Pomary–Center, 3,384 km, 30 CS;
Ureng–Pomary–Uzhgorod, 4,650 km, 42 CS.

The capacity of this very large gas transportation complex will be 200 billion cubic meters per year.

An additional 2500 kilometers and 12 compressor stations are planned for the Ureng–Pomary–Uzhgorod (string 2) gas pipeline and distributing pipelines along the primary mainlines.

The development of the network of gas mainlines and their length and capacity has a direct effect on the growth of the number and installed capacity of gas pumping aggregates (GPA) in compressor stations. This is demonstrated by the data in Table VI. In 1960, the number of compressor stations increased by more than a factor of 14 and the total installed GPA capacity increased by a factor of 68.4. The specific (per 1 km of pipeline) installed capacity of compressors changed during this period by approximately a factor of 11. Data on the change in the structure of the GPA stock are presented in Table VII.

On the whole, the state of the gas transportation system of the USSR at the end of 1981 was characterized by the following indicators:

length of gas mainlines, 135.5 thousand km;

fraction of large diameter (1020, 1220, and 1420 mm) gas pipelines, 59.5%.

Table IV Dynamics of the Length and Structure of Main Gas Pipelines

Years	Length, thousand km	Average yearly increment to the length of gas pipelines, km	Relative fraction of large-diameter gas pipelines, %		
			1020 mm	1220 mm	1420 mm
1960	20.9	4489	3.2	–	–
1965	42.3	5365	13.8	–	–
1970	66.0	4138	23.5	5.7	–
1975	98.7	7197	20.6	15.6	3.7
1976	103.0	4296	20.0	16.6	4.2
1977	111.3	8290	19.7	16.9	5.8
1978	117.7	6421	19.3	16.5	8.6
1979	124.4	6645	19.0	16.9	9.4
1980	131.6	7258	18.9	16.8	10.9

Table V Dynamics of the Increase in the Volumes of Gas Transported

Indicators	Years								
	1960	1965	1970	1975	1976	1977	1978	1979	1980
Volume of gas entering gas pipelines, billion cubic meters	26.8	107.0	179.1	272.7	302.2	324.8	348.2	377.8	403.1
Commercial gas, billion cubic meters	26.0	103.3	170.0	250.6	279.6	300.2	319.3	343.4	366.5
Rate of increase in the volume of gas entering gas pipelines, as a % of 1960 rate		399	668	1017	1128	1211	1299	1410	1504

Table VI Dynamics of the Increase in the number of and Installed Capacity of Compressor Stations

Years	Number of compressor stations	Number of compressor station aggregates	Total installed capacity of compressor station aggregates	Specific installed capacity of compressor station aggregates (for each km of gas pipeline)
	Units	Units		kw km^{-1}
1960	21	170	256.7	12.3
1965	81	665	1868	44.2
1970	130	1157	3400	51.5
1975	194	2093	8232	84.3
1976	214	2300	9603	93.2
1977	230	2518	11165	100.3
1978	254	2826	13488	114.6
1979	279	3084	15463	119.5
1980	303	3322	17571	133.5

Table VII Dynamics of the Change in the Structure of the Stock of Gas-Pumping Aggregates

Type of drive	Relative fraction of total installed capacity, %				
	1960	1965	1970	1975	1980
Gas-turbine	33	35	57	7.1	77.3
Electrical	37	53	30	19.4	17
Piston	30	12	13	9.6	5.7

volume of transported commercial gas, 419.2 billion cubic meters;

average distance over which gas is transported, 2167 km;

total installed capacity of gas pumping aggregates, 20.3 million kW;

consumption of gas for internal pipeline needs, 33.8 billion cubic meters per year.

Thus, over the last 20 years, the over-all length of gas mainlines has increased 6.2-fold, the volume of gas transported has increased 15-fold, and the power available per productive unit increased 68.4-fold.

These achievements are a result of scientific-technical progress in the gas industry, gained with the help of adjacent sectors of the national economy.

4.2. *Scientific-Technical Progress in Mainline Gas Transportation*

A characteristic feature of the development of the gas industry is the continuous improvement of the technology and the technical means of mainline gas transportation.

Scientific-technical progress in mainline gas transportation occurred along the following lines:

increase in the diameter of gas pipelines;

increase in the working pressure of the pipelines;

increase of the unit power of GPA with simultaneous improvement of their technical-economic indicators;

cooling the transported gas;

increase in the reliability of the gas supply system.

We shall examine each of these directions of scientific-technical progress.

Up to 1970, the principal means for increasing the efficiency of gas mainlines was to increase the pipe diameter. This indicator has the largest

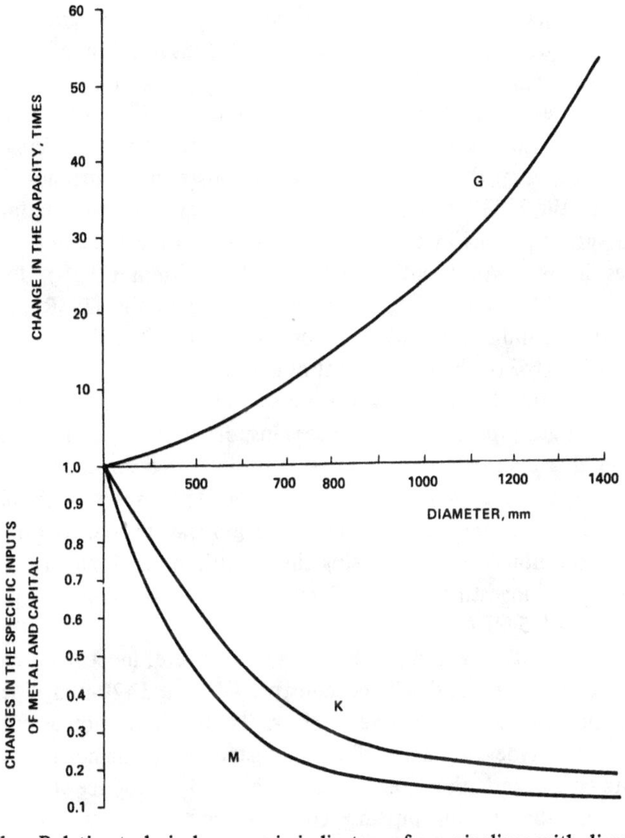

Figure 1 Relative technical-economic indicators of gas pipelines with diameters of 300–1400 mm.

effect of increasing pipeline capacity and improving all, without exception, technical-economic indicators of gas transportation. This is explained by the fact that as the pipe diameter increases, the throughput capacity of the pipeline increases in proportion to the change in diameter raised to the power 2.6, while the remaining indicators, such as metal consumption, volume of construction-assembly work, and, finally, capital investment, increase much more slowly than the capacity [13].

This is confirmed by the curves in Fig. 1, which reflect the change in the through-put (Q), specific metal input (M), and capital investment (C) as a function of pipe diameter. It is evident from Fig. 1 that when the pipe diameter increases from 300 to 1400 mm, throughput increases by more than a factor of 50.

One consequence of the sharp increase in pipeline throughput is lower specific (per unit mass of transported gas) consumption of metal and capital investment.

For the reasons mentioned above, the technical policy of the industry towards mainline gas transportation was directed toward increasing in every possible way the diameter of newly-constructed pipelines. The gas industry in the USSR assimilated the technology for constructing large-diameter gas pipelines in a very short period of time. The first main gas pipelines in the world with diameters of 1020 mm (1959), 1220 mm (1967), and 1420 mm (1972) were constructed in the USSR, In 1980, pipes with the indicated diameters constituted 46.6% and 1420-mm pipe constituted 10.6% of the total extent of gas mainlines.

It is interesting to note that in spite of the comparatively short extent of 1420-mm gas pipelines, these lines transport about 40% of all gas transported in this country. This is a result of the large throughput of these pipelines, which are pressurized at 7.4 MPa. The use of such pipe has made it possible to increase the volume of gas transported by a factor of 3.2 while simultaneously decreasing the specific metal input by 20% and specific capital investment by 34% compared to 1020-mm gas pipelines pressurized at 5.5 MPa.

At the present time and in the immediate future, most main gas pipelines are constructed and will be constructed with 1420-mm pipes [14]. Further increases in pipe diameter above 1420 mm are not useful at the present state of development of the industry due to higher pipe rigidity and buoyancy and the concomitant need to introduce fundamental changes into the existing pipeline construction technology and to create a special park of construction machines and machinery.

Under the conditions restricting further increases in pipeline diameter, the technical policy adopted by the industry toward increasing mainline capacity is to increase the pressure and decrease the temperature of the transported gas.

An increase in the working pressure of the mainline from 7.5 to 9.8 MPa is equivalent, with respect to increasing the throughput capacity and decreasing the costs, to increasing the pipe diameter from 1420 to 1620 mm.

The first step toward increasing the working pressure in gas mainlines was taken during the 9th Five Year Plan (1971–1975), when the fourth string of the Central-Asia–Center gas pipeline was constructed with 1420-mm pipes pressurized at 7.4 MPa. Before this, all pipelines in the country were designed for a standard pressure of 5.5 MPa. The increase in working pressure from 5.5 to 7.4 MPa increased the throughput capacity by 30–35% and decreased the capital investments and operational costs by 5–10%. However, the specific metal input remained practically constant at the previous level. This is explained by the fact that the throughout capacity and metal input change in direct proportion to the change in working pressure.

For this reason, improvements in the technical-economic indicators of gas transportation are achieved primarily by decreasing the costs of construction-assembly work. The results achieved and successful experience in operating the first high-pressure pipelines created favorable conditions for massive use of 1420-mm pipe for pipelines pressurized at 7.4 MPa.

All gas mainlines in the USSR are now constructed using the pipes indicated.

The next step of increasing the working pressure to 9.8 MPa will be taken during the current Five Year Plan by constructing the main pipeline from the Yamburg field in northern Tyumen province to the center of the country.

Technical-economic studies show that an increase in working pressure from 7.5 to 9.8 MPa will increase the capacity of the gas pipeline by 35% and will decrease the specific consumption of energy for gas transportation by 4–5%, while the specific capital investment and metal input will remain practically unchanged.

An important effect of an increase in pressure from 7.5 to 10 MPa is an overall decrease in the volume of construction-assembly work involved in the construction of multistring gas pipelines. According to total capacity, a three-string gas pipeline system pressurized at 9.8 MPa is equivalent to a four-string pipeline pressurized at 7.4 MPa. This circumstance, as well as the savings in fuel gas, is the decisive factor under the conditions of intensive growth in the volume of mainline construction and in the transportation distance.

The first step toward increasing the working pressure from 5.5 to 7.4 MPa was comparatively easily accomplished. However, the next step to 9.8 MPa involves complex problems, requiring the solution of technical problems in pipe production and assembly, based on the need for higher structural reliability of the pipeline.

The pipe quality, which determines the strength and plastic characteristics of pipes, has a decisive significance for increasing mainline reliability. Investigations of the reasons for and the nature of brittle fracture of pipes have established the requirements for the composition of the steel, the method for rolling sheet, and the technology for producing the pipes.

An important stage in improving the production of pipes for pipelines with high working pressure was the development of multilayered pipes, which permitted using the increased strength and plasticity of thin-rolled steel, developed by Soviet scientists under the leadership of Academician B. E. Paton, The possibility of increasing the guaranteed breaking point by 6–8% relative to thick sheet steel with the same chemical composition is entirely due to the use of a thin sheet.

But, the most important characteristic of multilayered pipes is their capability to attenuate crack propagation (the possibility of cascade pipeline failure is eliminated), as well as their lower sensitivity to surface defects compared to pipes with a continuous wall.

A test batch of 1420-mm pipes pressurized to 7.5 MPa was introduced in 1979. These pipes were used in the construction of a test section of gas pipeline. This permitted solving many problems in the technology of assembling multilayered pipes directly under the conditions existing along the pipeline route and testing the pipes for failure.

The studies performed confirmed the high carrying capacity and resistance of pipes to cascade failure. However, a number of problems involving an increase in the circumferential strength, longitudinal stability, transportability, and protection of multilayered pipes from corrosion require further investigations. The problem of ensuring longitudinal stability of a gas pipeline constructed from multilayered pipes must also be solved.

For realistic values of the temperature difference, determined by the conditions under which the gas pipeline is constructed and operated, longitudinal stability of the pipeline is ensured only along rectilinear sections and along sections with elastic bending in soils with natural moisture content at depths of one meter. In flooded soils and swamps, additional pipeline ballasting is required.

There is no doubt that effective technical solutions to the problems indicated will be found and multilayered pipes will be widely used in the construction of main gas pipelines pressurized at high working pressures.

Cooling the transported gas considerably increases the efficiency of gas mainlines.

At the present stage of development of the technology and technique for long-range transportation, gas cooling must be viewed not only as one of the methods for increasing the efficiency and reliability of pipeline systems, but also as a technologically required measure. The point is that as the diameter of gas pipelines increases, the effect of temperature factors increases and it is practically impossible to transport gas without cooling along 1220- and 1420-mm pipelines pressurized at a working pressure of 7.5 MPa.

Studies show that depending on the environmental and climatic conditions under which the pipelines are laid and operated and the achieved and predicted levels of development of the production of refrigeration equipment, it is useful to separate the cooling of transported gas into two levels: cooling to temperatures close to the ambient air temperature and moderate cooling to temperatures close to the temperature of the soil at the depth at which pipelines are laid in northern regions.

It is useful to cool the gas to ambient air temperatures on pipelines with a diameter of 1020 mm and larger under all environmental and climatic conditions. Air cooling equipment (ACE) is most widely used to cool gas to this level.

Technical-economic studies show that the maximum effect from the use of ACE is achieved when cooling gas with yearly average conditions to a level exceeding the ambient air temperature by 8–12° C.

The use of ACE on main gas pipelines with diameters of 1220–1420 mm increases the annual pipeline productivity by 6–8% and decreases the specific reduced cost by 3–4%.

It is also significant that ACE provides the greatest depth of cooling of gas and, therefore, maximum pipeline capacity during the winter (the period of maximum gas utilization) as well.

In the immediate future, a significant part of newly constructed gas pipelines will pass through the northern regions of Western Siberia, which are characterized by the presence of unstable and permafrost soils with difficult geocryological conditions.

Construction of gas pipelines under these conditions makes it necessary to solve a number of complicated technical problems, one of which is to eliminate the thermal interaction between the gas pipeline and the surrounding medium (permafrost).

Based on the results of studies and the analysis of climatological data for the regions in which the gas pipelines are to be constructed, the optimum depth of cooling of gas was set at minus 1–4° C.

Industry institutes are working on the problem of deeper cooling of gas to temperatures close to the critical temperature in order to transport gas in the cooled and liquified states.

It is possible that the technology of transporting gas in the cooled gaseous state at a temperature of minus 65–70° C and a presure of 10–12 MPa will permit decreasing the specific metal input by a factor of 1.8–1.9 and specific energy expenditures by 25%. The specific metal input can be decreased by another factor of two when transporting gas in the liquified state by decreasing the optimum working pressure to 5.0 MPa with practically equal energy expenditures as compared to existing technology for pumping gas at the usual temperatures.

Realization of the gas transportation technologies indicated above involves the solution of a complex of complicated scientific-technical problems, beginning with the development of new materials and equipment for low temperature gas pipelines and the technology and technical means for constructing pipelines under difficult route conditions. For this reason, the practical use of low-temperature gas transportation technology must be viewed in terms of the distant future.

The technical policy of increasing the capacity of gas pumping aggregates and compressor stations as a whole was put into effect simultaneously with the concentration of gas flows.

Based on an analysis of technological and technical-economic indicators, as well as optimum intervals of variation of the productivity of main gas pipelines, industry institutes provided a justification for the use of gas pumping aggregates with the following series of parameters: 4–6–10–16–25 MW.

Up to the present time, GPA with capacities of 4.6 and 10 MW have been constructed and successfully operated, and new aggregates with a capacity of 16 and 25 MW are in the initial stages of serial production in this country.

The main indicators of assimilation of the production of domestic GPA with gas turbine drive are presented in Table VIII. In addition to the aggregates indicated, new modifications of GPA are planned for series production for the following:

booster compressor stations operating at supercharge pressures of 4.0, 5.5, and 7.4 MPa driven by 6.3- and 16-MW gas turbine plants (GTP) of the type used in aircraft engines;

pumping of corrosive gases with 6.3- and 10-MW GTP.

The realization of the series of GPA indicated above with full-head superchargers with a replaceable flow-through part will lead to an optimum standard size for compressor stations for gas pipelines with different diameters and it will ensure high loading and efficient utilization of equipment during the entire operational lifetime. For 1420-nm pipelines pressurized at 7.4 and 9.8 MPa, compressor stations with 16- and 25-MW aggregates are optimum. Calculations show that compressor stations with the aggregates indicated will be characterized by lower inputs: capital investment lower by a factor of 1.5–2.5, labor costs lower by a factor of 1.6–2.4, and metal consumption lower by a factor of 2.8–3 compared to analogous compressor stations equipped with 10-MW GPA. This predetermines the predominant use of 16- and 25-MW GPA on newly constructed main gas pipelines [15]. In the next period of development of the unified

Table VIII Basic Indicators of Gas-pumping Aggregates with Gas-turbine Drive

Name of indicator	GTP-750-6	GTP-6-750/GTP-6	GPA-Ts-6.3	GTP-10	GPA-10	GTP-16	GTP-16	GPA-Ts-16	GPS-25
Manufacturer	Nevskii works	Turbomotornyĭ works	Frunze Construction-Assembly works	Nevskii works	Yuzhnyĭ turbinnyĭ works	Turbomotornyĭ works	Turbomotornyĭ works	Frunze Construction-Assembly works	Nevskii works
Type of drive	stationary with regeneration	stationary	aircraft	stationary with regeneration	marine	stationary	stationary	aircraft	stationary
Power, thousand kW	6.0	6.0	6.3	10.4	10.0	16.0	16.0	16.0	25.0
Effective efficiency of gas-turbine, plants under conditions in the field, %	27.0	24.0	22.5	29.0	26.5	25.0	29.0	27.5	27.5
Temperature of the cycle, °C	760	760	710	780	785	810	920	794	920
Capacity of super-charger, million m day^{-1}	20.0	20.0	10.7	36.0	36.0	51.0	31.0	33.25	63.0
Degree of compression by supercharger	1.24	− 1.24	1.45	1.23	1.23	1.24	1.44	1.44	1.44
Specific metal input into the gas-pumping aggregate in the factory production, kg kW^{-1}	13.7	11.6/13.8	11.6	15.7	5.6	9.4	7.5	10.6	6.3

gas supply system, the relative fraction of 16- and 25-MW GPA will continue to increase.

GPA with full-head centrifugal superchargers will be widely used. Introduction of such aggregates will increase the reliability and productivity of gas pipelines, decrease the requirements imposed on GPA, simplify the technological design of compressor stations, and decrease the energy expended on compression.

Full-head superchargers are graded according to the replaceable flow-through part based on the degree of compression 1.35–1.37, 1.45, 1.52–1.55. The use of a replaceable flow-through part permits efficient use of drive power under conditions when there is an unavoidable spread in the distance between compressor stations and large amounts of gas are removed along the pipeline route.

Under conditions of intensive construction of compressor stations, solutions to technical problems that decrease the volume of construction-assembly work performed on location, decrease total capital investments, and permit easy and rapid replacement of failed parts by unit and aggregate repair of GPA are of prime importance [16, 17]. New GPA with gas turbine transport engines best satisfy the requirements. At the present time, two types of such aggregates are used in the industry: GPA-Ts-6.3 driven with a 6.3-MW aircaft type GTP and GPA-10 driven with a 10-MW marine GTP.

Of the new designs, aggregates of this type include GPA-Ts-16, GPS-Ts-25, and GPA-Ts-10. These block, basementless automated aggregates are designed to operate under different climatic conditions. They are placed in factory-made containers or in easily erected individual buildings and, together with other technological equipment, they make it possible to use industrial methods for constructing compressor stations. Comparatively easy conditions for dismantling and transporting engines and other GPA permit complete utilization of the advantages of performing maintenance work under factory conditions. The newly developed plants have quite high efficiencies: 26% for GPA-Ts-10, 27% for GPA-Ts-16, and 32–33% for GPA-Ta-25.

It should be noted that due to forcing of the parameters of its thermodynamic cycle, the GPA-Ts-16 plant is based on the NK-12 ST drive used in GPA-Ts-6.3. The remaining equipment of the GPA, with the exception of the flow-through part of the supercharger, also remains unchanged. This reconstruction of the plant permits future modernization without special difficulties of the existing GPA-Ts-6.3 park.

The modernization indicated will decrease the utilization of gas for the internal needs of compressor stations approximately by 1.5 billion cubic meters/year.

The problem of conserving fuel-energy resources in mainline gas transportation systems becomes urgent in connection with the increasing distance of gas transportation and increasing input-to-output power ratio of gas pipelines.

One way to decrease gas consumption for the internal needs of compressor stations is to increase the energy efficiency of the compressor equipment used.

Improvement of the power equipment (introduction of more powerful and efficient plants) over the last ten years has increased the efficiency (decreased the specific consumption of fuel) of the park by 7%. This trend will continue in the future.

One possible solution of the problem stated could be the development of the regenerative variant of GTA with an efficiency of about 33–35%. Of course, the use of the regeneration cycle complicates the construction of the plant, which violates the requirements formulated above for industrialization of the construction of compressor stations. For this reason, here, it is necessary to find a solution which satisfies as closely as possible the requirement for increasing the efficiency of GPA while maintaining industrial methods for assembly of compressor stations.

From here it follows that the most useful way to increase the economic practicability of gas turbine installations while simultaneously decreasing the specific (per unit power) mass is to increase the initial parameters of working GPA processes (temperature, pressure, and speed), based on overall progress in the construction of turbines.

The use of high-power combined steam-gas cycles, characterized by high (40–45%) efficiencies, in GPA is particularly promising.

Significant decreases in utilization of gas for internal needs involve extensive utilization of electrical drive at compressor stations.

In recent years, the relative fraction of electric drive in the gas industry decreased to 17%. The decrease resulted from the fact that most of the large gas pipelines were constructed far away from power generating systems and from the lack of large electric-drive plants.

The present trends in the evolution of the energy balance of the country, which are related to the increased role of gas as a specialized and mobile fuel, as well as the development of nuclear power, once again create objective conditions for wide application of electric drive, especially in interconnected gas transport systems in the center of the country.

It is proposed that the fraction of electric drive be increased by increasing the generation of electricity by nuclear power plants and thermal electric power plants, burning coal with low calorific value.

Electrically driven compressor stations have a number of advantages over gas turbine stations, the most important of which are lower (by 35–40%) capital investment in the construction of the compressor station, higher reliability, and lower danger of fire and explosion. Electric drive is also promising on the economic and social level in connection with environmental protection, as well as for providing high levels of automation and creation of comfortable living and working conditions for service personnel at compressor stations.

Summarizing what was said above, it should be noted that the parameters of main gas pipelines, compressor stations, and gas supply systems as a whole change in response to the specific conditions of development of the gas industry and scientific-technical progress in other sectors of the economy such as metallurgy, machine building, thermal power, instrument building, etc.

4.3. Organization of the Unified Gas-Supply System

Together with the increase in the length and capacity of gas mainlines, expansion of the geographical distribution of gas and gas-condensate fields brought into production, the gas transportation subsector of the gas industry of the USSR has also undergone structural changes, related to the formation of multistring mainline systems and the construction of intersystem ties and circuits around large industrial centers. These changes were caused by the need to maneuver large flows of gas in order to increase the reliability of gas supply to consumers, spread out over an enormous territory.

The first step in this direction was taken at the beginning of the 1960s with the construction of multistring main gas pipelines from the Northern Caucasus, the Ukraine, and Central Asia into the central regions of the country and the Urals. In later years, main pipelines were designed with regard of the presence of a branching network of gas mains and intersystem ties between them, as well as the plan for creating the unified gas supply system of the country.

Thus began the formation of the UGSS, which is continuing at the present time with new mainlines and underground storage being added to the existing system.

In the Soviet Union, the following systems presently exist and are in operation: Central, Ukrainian, Western, Povolozh'e, Central Asian, Ural, Trans-Caucasian as well as Central-Asia–Center, and northern Tyumen province to the Center. Their interaction and mutual backup provide flexibility in maneuvering gas flows and increase the reliability of gas supply to the industrial centers of the country due to compensation of decreased gas delivery from some gas producing regions by increased delivery of gas from other regions.

Thus, for example, the high-capacity Northern-Caucasus–Center and Bukhara–Ural gas transportation systems, intended for transporting gas from north to south, are beginning to function as intersystem ties of rapidly developing gas transportation systems transporting gas from northern Tyumen province to the central regions of the European part of the USSR. This indicates the enormous possibilities inherent in the unified gas supply system of the country not only for maneuvering large flows of gas, but also for maximum utilization of the capacities of gas production and transportation enterprises.

The internal turnaround as fraction of the total volume of gas entering mainlines is an indirect indicator of the unification of gas transportation systems into a unified gas supply system.

Internal turnaround is defined as a bypass of part or all of the gas from one gas transport system into another in order to ensure uninterrupted supply of gas to consumers while emergency or preventive maintenance is performed on gas pipelines and compressor stations, as well as to cover the load due to underground gas storage in different regions of the country.

Table IX presents data on the volume of gas transported along the unified gas supply system of the USSR including internal turnaround for the period from 1960 to 1980.

As is evident from Table IX, the relative fraction of internal turnaround in the total volume of gas entering the pipeline is constantly increasing. In 1960, it constituted about 15% and in 1980 the overflow fraction exceeded 66%.

The increasing trend in the indicator of internal turnround in the unified gas supply system will remain in the future. This imposes additional requirements on the design of new gas main pipelines, which must take into account the coupling of the system with transport, storage, and gas distribution objects.

Compared with other sectors of the fuel-energy complex of the country, UGSS has a number of particular characteristics. The main features of UGSS are its large scale, dynamism, and high concentration of production capacity.

The dynamism of the expansion of UGSS is related with the high rates of development of the gas industry and the systematic increase in the contribution of gas to the energy balance of the country, mentioned above. The high concentration of UGSS capacities is expressed in terms of the increase in gas flows due to the use of larger diameter pipelines

Table IX Internal Turnaround as a Fraction of the Total Volume of Gas Entering Gas Pipelines

Name of indicator	Years				
	1960	1965	1970	1975	1980
Gas entering pipeline including internal turnaround, billion m^3	31.6	158.8	288.7	584.7	1197.2
Internal turnaround, billion m^3	4.8	51.7	119.6	311.9	794.1
Internal turnaround as a percent of the volume of gas entering pipeline, %	15.2	32.6	40.0	53.3	66.3

(1220–1420 mm), increase in the pressure used in mainline gas transportation (in the future up to 9.8 MPa), and the use of gas pumping plants with high unit capacity (up to 25 MW).

The UGSS is developing within the framework of the entire fuel-energy complex of the country. In this connection, in optimizing the structure of UGSS, it is necessary to change over from local optimization of separate elements to optimization of some sector singled out of the economy, which, aside from the gas supply system, includes a number of its consumers and suppliers, as well as suppliers of alternative energy resources. In addition, it is very important to include uncertainty and instability of development of the UGSS; this gives rise to the creation of a flexible gas supply structure, capable of adapting to changing conditions. The possible changes in the conditions of development of UGSS concern the security of gas resources, scope and periods of introduction of new technology, and so on. In many cases, the sizes and periods of consumption of special purpose installations change. Changes in the status of the sector supplying the gas industry with pipe, gas pumping equipment, and so on have a large impact. An efficient way to ensure reliable and stable development of UGSS is to create a flexible gas transportation network, capable of adapting to the conditions of the period in question with minimum costs. It should be emphasized that the value of the flexibility factor increases constantly. In other words, a change in the external conditions of development of UGSS leads to increasingly more significant consequences in the level of gas supply to consumers.

Future development of UGSS will be determined mainly by increased gas production in northern Tyumen province. One of the principal features of plans for the future development of UGSS is providing for systems properties of the plan that conform to the requirements of not only economic practicability and efficiency, but also reliability (including viability) of gas supply. Among the different measures ensuring operational reliability and flexibility of UGSS, together with traditional methods (increasing reliability of pipes and equipment, backup plants at compressor stations, construction of ties between pipe strings, etc.), systems methods for ensuring network-structural reliability of the system and interaction of different subsystems of the UGSS (gas-fields, main gas pipelines, underground gas storage) are becoming increasingly more urgent.

To ensure network-structural reliability of UGSS, the requirements for viability and flexibility of the system, i.e., its capability to withstand strong perturbations, involving, for example, a sharp decrease in the capacity of separate large pipelines, should be kept in mind. Starting from this, provisions for transporting Tyumen gas to regions where it is used are made along several directions: central, northern, and southern. The reliability of gas transportation and the stability of its delivery to consumers is ensured by comprehensive solutions to technical problems, provided for in the construction of separate gas pipelines. Each large gas pipeline designed is viewed as a stage in the development of UGSS as a whole and its reliability is ensured by taking into account the interaction with other objects in the system. This forms the basis on which the size and distribution of reserves in underground gas storage (UGS), the nodes at which each gas pipeline is connected to the UGSS for efficient maneuvering of gas flows, the magnitude of the production capacity of gas fields, as well as the volumes and placement of production capacities of service and repair systems, are determined.

The optimal backup in the gas supply system must be determined by the following solutions to technical problems:

creation of backup for gas production and gas fields, which is primarily secured by drilling additional operational wells;

backup of the intrinsic gas pipeline by installing an efficient backup capacity of pumping plants at compressor stations, construction of ties between parallel gas pipeline strings, and main ties between separate gas pipeline routes:

creation of operational gas reserves in underground storage and expansion of the UGS network.

In conclusion, we note that the development of the unified gas supply system is in many ways analogous to the development of the unified electrical power system of the country [18]. Both systems consist of the same technological elements: sources, transmission lines, consumers, and production regulators. Both systems started out as separate, unconnected systems.

Technical progress in UGSS and the unified electrical power system is also identical and is related with high concentration of production and increasing capacity and productivity of technological equipment.

5. Underground Gas Storage

The close technological relation between the elements of the UGSS, beginning with gas fields and ending with consumers, raises the problem of creating storage (accumulators) for gas, which could compensate (smoothe) planned and unplanned disturbances of the system.

Planned disturbances include seasonal and daily nonuniformities in gas utilization; unplanned disturbances include different types of accidents in the gas- and fuel supply system. Unplanned disturbances also include an anomalous component of the seasonal nonuniformity in gas consumption, caused by sharp of prolonged drops in temperature during separate winter periods.

The large volumes of gas transported, the large distances between the sources of gas and the centers where the gas is consumed, and the possibility of the appearance of large disturbances in the gas supply system make it necessary to create an extended network for storing gas.

A certain volume of gas, which could enter into the gas supply system during planned and unplanned disturbances, must be accumulated in the storage facilities. The gas volume that compensates for planned disturbances must be set aside for seasonal extraction during the winter period. To compensate for unplanned disturbances, an operational gas reserve must be stored underground to ensure reliable functioning of the UGSS.

Accordingly, the UGSS must include base storage facilities, intended to provide for planned disturbances in the flow of gas as well as "peak" storage to cover unplanned disturbances.

As a rule, underground storage in depleted gas fields and in water-bearing formations serves as base storage. Storage in salt formations and storage of liquified natural or petroleum gas can serve as "peak" storage.

The presence of a network of "peak" stored reserves will increase the role of the gas supply system in regulating the daily nonuniformity in the utilization of fuel and energy in the national economy.

The first underground storage facility in the USSR was created in 1957. Commercial pumping of gas into the first underground storage facility constructed in the USSR in a water-bearing formation began in 1959.

The USSR now has a large number of underground storage facilities. The series of underground storage facilities represents the largest such facility in the world.

During the initial period of development of underground gas storage in the USSR, a great deal of attention was directed toward creating storage in water-bearing structures. As the unified gas supply system of the country developed, it became economically feasible to create large (base) storage in depleted gas fields.

At the present time, the role of underground gas storage has increased even more in connection with further increases in the relative contribution of gas to the energy balance of the country and in the distances between the principal source of gas and the large industrial centers.

Plans have been developed for further development of underground gas storage, especially in the gas consuming regions of the country. The problem of creating underground gas storage along main gas pipeline routes, which will increase the reliability of the gas supply system as well as its yearly capacity, is being examined at the same time.

Increasingly more difficult extraction-geological conditions of gas production in a number of regions of the country (increase in well depth, increase in the fraction of fields with a high content of condensate and corrosive components in the gas) limit the possibility for regulating the seasonal nonuniformities by means of regulator–fields and raises the problem of creating underground gas storage in depleted fields in these regions.

It is well known that optimal conditions for smoothing nonuniformities in the utilization of gas occur with a definite combination of base storage facilities in depleted fields, situated in direct proximity to large gas transportation systems and regulating the seasonal nonuniformity of one or a series of large industrial regions with relatively small ("peak") storage, intended for providing additional gas supply during a period of peak demand. The largest underground storage facilities, created by leaching in salt formations, could successfully fulfill the role of peak gas storage. This type of underground storage is currently in use in the USSR.

Concrete experience in searching for small structures (10–15 m) has been accumulated in this country. The suitability of water-bearing formations for gas storage is determined by geometrical, hypsometric, geological-geophysical, hydrogeological, and hydro-gas-dynamic parameters of both the space into which gas is pumped and of the cap, and including the deposits lying above the cap. The parameters listed above cannot be established by any one method. A combination of different methods for studying the parameters of the formation must be used. The most complicated and important problem of exploratory work is the study of geological-physical properties of the formation and the quality of the seal provided by the cap.

The method of experimental-commercial exploitation of storage is widely used in the USSR. This period, which marks the beginning of the operation of pumping gas into the storage volume, is a logical continuation of exploratory work on this volume. Experimental-commercial exploitation of the storage formation makes it possible to obtain more accurate values of the geological-physical parameters of both the formation and of the cap and to make a more reliable determination of the technological characteristics of future storage volumes.

Hydrodynamic methods of investigating formations and wells are widely used in exploring water-bearing structures. The extensive use of these methods is a result of the necessity of making more complete use of the structure explored, as well as the necessity of obtaining starting information for further planning. A complex of hydrodynamic methods has been devleoped for studying formations and wells, including testing of wells for steady-state and non-steady-state regimes, investigation of wells using the interaction method, as well as the method of pneumo- and gas-dynamic exploration. Methods have been developed for analyzing the results obtained from investigations of wells, which permit determining the percolation-capacitance properties of the formations under different geological conditions and methods for exciting the formation. In recent years, methods for investigating formations and wells and for analyzing results of measurements, based on the use of the theory of optimal control and on the extensive use of computational techniques, have been used.

Hydroexploration methods provide quite reliable information on the formation-reservoir and on the rocks that form the cap of the future storage volume. The percolation-capacitive properties of the formation, the productivity of the wells, the maximum permissible depression in the formation, the direction of predominant motion of gas, the measures for intensifying the flow of liquid and gas into the well, the quality of the seal provided by the cap, and other factors are determined.

Reliable starting information on the formation-reservoir allows for highly accurate forecasts of future creation and exploitation of underground storage. For this purpose, the scientific foundations of hydro-gas-dynamic methods for calculating underground gas storage, which permit calculations of the process of creating the storage under different geological conditions, including in formations that are highly nonuniform, have been developed.

Based on the theory of quasi-steady-state fluid and gas flow in a porous medium, a theory has been developed for cyclic exploitation of the storage formation, when the pumping volume approximately equals the volume of gas removed from the formation. Computer calculations are being performed using data obtained from investigations of formations and wells, as well as the results of experimental investigations. Subsequent pumping and removal of gas, as a rule, confirm the predictions of hydro-gas-dynamic calculations.

The requirement of maximum utilization of the explored structures stems from the need to increase the pumping pressure, which increases the volume of reactive gas and decreases the time required to create the UGS facility.

Calculations show that the pressure in the formation after gas is pumped in can greatly exceed the initial hydrostatic pressure in the formation (by up to 50–70%); in addition, the seal provided by the cap, formed by clay deposits, remains intact. The main problem here consists of ensuring that the wells are sealed. For this purpose, methods have been developed to monitor the sealing of the wells used in UGS, and special technical requirements, ensuring reliable operation of the wells, have been developed for the construction of wells used in underground storage.

From the very beginning of the development of underground storage, there arose the problem of preventing the collapse of the face of the formation, formed by uncemented or weakly cemented rocks. To prevent collapse of the formation-reservoir and the removal of sand, different types of filters with different construction are placed in the filtering zones of the wells. The most widely used filters are gravel-wash and suspended (frame-wire, plexiglass, metalloceramic, titanium-magnesium, monodacron, etc.) filters.

The type of filter and the technology used to install it are determined by the specific geological-field conditions.

Experimental work is being performed on investigating the process of collapse of the face when gas or fluid moves in the formation.

Special attention was directed toward developing methods for performing calculations of underground storage reservoirs created in low rises. The problem was to determine the movement of the gas pumped in along the roof of the formation, the rate at which gas should be pumped into the formation so as to ensure that the trap is utilized to fullest extent possible and that the storage reservoirs are created within fixed periods of time, and the distribution of wells and the nature of the section of the formation-reservoir. Criteria have been established for determining gas motion in low rises and the degree to which they affect the utilization factor of the trap. In a number of storage reservoirs, gas is forcibly pumped into the lower intercalation in order to increase the effective volume of gas in low rises. Particular experience has been accumulated indicating the utility of pumping gas into the lower intercalation, separated from the main gas pool by deposits with poor reservoir properties.

An underground storage reservoir for gas in a horizontal formation, which is unique in the world, has been created in the northwestern part of the Soviet Union. The calculations and experimental investigations performed established that gravity-induced flow of gas into the water-bearing formation is insignificant. This provided the theoretical basis for subsequent pumping of gas into a formation without a structural form. Subsequent test operations on this underground gas storage reservoir basically confirmed the theoretical assumptions.

A number of measures are being taken to decrease flow of gas out of the gas-bearing volume and to decrease the flow of water into the wells in the UGS in a horizontal formation, including pumping water into special wells situated beyond the gas-saturated part of the formation in the zone of monoclinic uplift of the formation, displacing the line along which gas is injected into the formation, decreasing the neutral period, and pumping surfactants into the formation.

To calculate the process of pumping gas into gently-sloping and horizontal formations, experimental investigations of the displacement of water with gas were performed. The dependence of the coefficient of displacement of water with gas on geological and technological factors was established. Experimental investigations of the influx of water into the UGS accompanying the removal of gas were also performed.

The requirements for maximum utilization of geological storage facilities (providing high pressures, pumping maximum permissable amounts of gas into the formation) lead to the necessity of detailed monitoring and observation of the development and operation of the UGS facility. Monitoring and observation are performed by a complex of methods (gas-dynamic, hydrodynamic, hydrochemical, geological, geophysical, gasometrical, and others). These operations are performed in operational wells as well as in wells specially chosen for this purpose (observational, piezometrical, monitoring). The network of observational and monitoring wells is selected so as to obtain complete and reliable information on the formation into which gas is pumped as well as on the upper control horizons.

Scientific-research institutes have accumulated a great deal of experience in analyzing the technical-economic indicators of underground storage reservoirs for gas. Methods based on dynamic programming have been developed for calculating the optimum placement of underground storage reservoirs in the gas supply system of the country, which provides the required volume of reactive gas for controlling seasonal nonuniformaties.

Computer programs have been developed for calculating the basic technological parameters of the storage reservoirs. These parameters permit determining the optimum relation between the reactive and buffer gas, the number of operating wells, and their distribution.

6. Gas Refining

Gas refining, as the most important sector of the gas industry, began to develop intensively in the middle of the 1970s. This was facilitated by the opening of large gas fields, whose gas contained, aside from methane, such valuable components as ethane, propane, and heavier homologs, as well as hydrogen sulfide and helium in commercial quantities. An example of such a field is the Orenburg field. The largest gas refining complex in the country was developed based on this field [1, 19]. At the present time, the Orenburg gas refining complex annually produces scrubbed natural gas, sulfur, stable condensates, a wide fraction of light hydrocarbons, and liquified gases. Absorption technology utilizing diethylamine as the absorbent is used to scrub the gas. This technology provides a high degree of cleaning of the gas, down to a H_2S content of less than 5.7 mg m^{-3}, as well as the required amount of acidic gases for their further refining into sulfur using Klaus' method [20]. The capacity of a single scrubbing installation can be as high as 8 million cubic meters of gas per day (per single absorber) and that of Klaus' installations can be as high as 1000 tons of sulfur per day. At the same time that the Orenburg project was being realized, work was being conducted on another large gas refining complex: the Mubarek complex (in Central Asia). The complex is completely equipped with domestic equipment and is calculated to refine low-sulfur (less than 0.3% by volume of H_2S) and sulfurous (5.5% by volume of H_2S) natural gas. The gas is likewise scrubbed by diethylamine and sulfur is obtained by Claus' method. The unit capacity of the absorber in the scrubbing installation for low-sulfur gas is 6 million cubic meters per day and the unit capacity of the sulfurous gas installation is 4.5 million cubic meters per day.

The commercial product produced by the Mubarek complex is scrubbed gas, sulfur, and stable condensate.

In constructing the Orenburg and Mubarek gas refining complexes, a number of complicated technical problems were solved, such as creating equipment with high unit capacity, assimilating the technology for deep extraction of valuable components from sulfurous gas, and protecting the environment.

Together with the refineries indicated above, the industry is successfully operating the Shebelinsk, Ukhtin, Azerbaĭdzhan, and Moscow gas refineries as well as a number of plants for stabilizing condensates.

Work has now begun on creating the Astrakhan gas-chemical complex (GCC). The high content of H_2S in the formation gas (25% by volume) in this field makes this field a large source of gaseous sulfur.

A gas refining complex based on the Shurtan field (Central Asia) is also being developed. The formation gas of this field contains hydrogen sulfide (0.1% by volume) and condensate. Absorption technology is used for removing hydrogen sulfide from the gas.

An important part of natural-gas refining is the production of gas black. Furnace black and thermal black are produced from natural gas primarily to meet the requirements of the tire and rubber industries. Experimental and theoretical investigations of soot formation accompanying thermal decomposition and combustion of hydrocarbons are being conducted. The results of these investigations are important both for the technology of soot production and for protecting the environment from the emission of soot as a by-product of combustion.

In the future, the gas refining subsector will grow through the development of large refinery complexes as well as through complex refining of gas which makes use of all valuable components in the gas. The gas industry is thus becoming not only a supplier of fuel, but also a large producer of an entire series of valuable products for the national economy.

A very valuable hydrocarbon raw material for the chemical and petrochemical industries is stable gas condensate. World practice shows that the cost of producing gas condensate is approximately two to four times lower than the cost of producing oil, while specialized refining of condensate at special plants increases the yield of white products approximately by a factor of 1.5 above the yield obtained in primary refining of petroleum.

At the present time, most of the gas condensate in the USSR is refined in petroleum refineries together with petroleum. In most cases, gas condensate is stabilized at gas refineries and some of it is refined for the production of motor fuel, solvents, and other chemical products.

Different brands of motor fuel are also used in gas production and in construction operations by the gas industry together with gas itself.

Taking into account the fact that transporting motor fuel to separate regions of the country involves large expenditures of resources, it is now recognized that it is expedient to refine gas condensate on location in volumes sufficient for local requirements. Small block setups are used for these purposes. These setups can also be successfully used in gas fields with low production of gas condensate, which is difficult to transport. Such setups are currently being used in Western Siberia.

Gas condensates from different fields differ considerably in their fractional and hydrocarbon composition [20]. For this reason, small setups, depending on the physicochemical properties of the condensate and assortment of products obtained, are assembled from several basic technological blocks.

Further development of the gas refining subsector will involve the creation of large gas-chemical complexes, using natural gas from fields in Western Siberia, Komi ASSR, Central Asia, and the Astrakhan province with a high content of ethane, propane, butane, gas condensate, and acidic components as a raw material.

The outlook for creating GCC in these regions is determined by the large reserves of gas containing commercial quantities of the target components. To increase the profitability of the gas refineries situated in unfavorable geographical regions, solutions are being developed to the technical problems involved in increasing the level of automation of the installations as well as in creating fully-equipped–block equipment.

To simplify the technological process involved in refining natural gas containing sulfurous components, ethane, and condensate, complex processes utilizing selective physical absorbers are being developed.

The principal products of natural gas refineries at GCC will be ethane and liquified gases. Ethane and the heavier hydrocarbons will be extracted using a non-energy-intensive, low-temperature technology based on the vapor compression cycle with a multicomponent coolant and turbo-expansion engines for refrigeration.

An alternative method for extracting C_{2+} hydrocarbons directly at the gas-condensate field is being developed simultaneously as a part of the plants engaged in complex preparation of gas and condensate for transportation. The use of plants for deep refining of gas at large gas condensate fields in the northern part of the country will eliminate the need to create an independent system for collection, treatment, and transportation of gases containing ethane.

References

1. Orudzhev, S. A. (1976). *Gazovaya promyshlennost' po puti progressa (The Gas Industry on the Path to Progress)*, Nedra, Moscow.
2. *Materialy XXVI s"ezda KPSS (Information from the 26-th Congress of the Communist Party of the Soviet Union)*, Politizdat, Moscow (1981).

3. *Gazovye i gazokondensatnye mestorozhdeniya. Spravochnik (Handbook on Gas and Gas-Condensate Fields)*, Nedra, Moscow (1975).
4. Orudzhev, S. A. (1981). *Goluboe zoloto Zapadnoĭ Sibiri (Blue Gold of Western Siberia)*, Nedra, Moscow.
5. Margulov, R. D., N. I. Belyĭ, A. I. Gritsenko, and T. P. Shmyglya (1982). *Complex Assimilation of Gas Fields in the Northern Regions of the USSR*, Report at the 15th International Gas Congress, Lausanne, Geneva.
6. Gritsenko, A. I., V. I. Ermakov, I. P. Zhabrev, G. A. Zotov, and V. P. Stupakov (1982). *Long-Term Forecasts of the Levels of Gas Production, Gasovaya Promyshlennost'*, No. 10, 8–10.
7. Margulov, R. D., E. K. Selekhova, and I. Ya. Furman (1976). *Razvitie gazovoĭ promyshlennosti i analiz tekhniko-ékonomicheskikh pokazateleĭ (Development of the Gas Industry and Analysis of Technical-Economic Indicators)*, VNIIÉgazprom, Moscow.
8. Sherbinina, B. E., Yu. I. Bokserman, A. D. Sedykh, *et al.* (1981). *Otechestvennyĭ transport (Domestic Pipeline Transportation)*, Nedra, Moscow
9. Kortunov, A. K. (1967). *Gazovaya promyshlennost' SSSR (The Gas Industry in the USSR)*, Nedra, Moscow.
10. Sorokin, A. I., S. R. Derezhov, and V. A. Mysyakin (1982). *Experience in Design and Construction of the 'Soyuz' Gas Pipeline*, Report at the 15-th International Gas Congress, Lausanne, Geneva.
11. *Gas – Truby. Rasskaz o tsentral'noĭ stroĭke pyatiletki (Gas – Pipes. The Story of the Heart of the Five-Year Plan)*, Izvestiya, Moscow (1982).
12. Dinkov, V. A. (1982). *To Serve People Reliably*, Pravda, No. 277, October 4.
13. Khadanovich, I. E. *Analiticheskie osnovy proektirovaniya i ékspluatatsii magistral'nykh gazoprovodov (Analytical Foundations of the Design and Operation of Main Gas Pipelines)*, Gostoptekhizdat, Moscow.
14. Dinkov, V. A. (1982). *Achievements of and Prospects for Further Development of the Gas Industry in the USSR*, Report at the 15th International Gas Congress, Lausanne, Geneva.
15. Kuznetsova, A. L., V. E. Evdokimov, Yu. G. Korsov, and P. V. Khrabrov (1980). *Advantages of Enlarging the Unit Capacity of Gas-Pumping Aggregates*, Énergomachinostroenie, No. 2.
16. Vasil'ev, Yu. N. (1977). *Gas-Transportation Equipment for the Industrial Method of Construction of Compressor Stations*, Gazovaya promyshlennost', No. 4.
17. Shuran, N. V. (1982). *Standard Designs of Unified Compressor Stations with Different Types of Gas-Pumping Aggregates*, Gazovaya promyshlennost', No. 9.
18. *Soviet Technology Reviews, Sec. A, Energy Reviews, Nuclear Power Systems*, Gordon and Breach, New York (1982), Vol. 1.
19. Mishin, V. M. (1981). *Problems for the 11th Five Year Plan in the Preparation and Refining of Natural Gas, Podgotovka i pererabotka prirodnogo gaza*, VNIIÉgazprom, Moscow, No. 1.
20. Cowle, A. P., and F. S. Rizenfeld (1978). *Ochistka gaza (Scrubbing of Gas)*, Translation from English, Nedra, Moscow.

Sov. Tech. Rev. A Energy Reviews, Vol. 2, 1985, pp. 155–171
0275–7893/85/002–155 $30.00/0

FORCED DEVELOPMENT OF GAS FIELDS

S. N. ZAKIROV and Yu. P. KOROTAEV

Moscow Institute of Petrochemical and Gas Industry
117917, Moscow, Leninsky Bv., 65

The growth of the gas industry in the USSR is marked by real achievements. Further intensive development of the gas production industry will require solutions to many complex problems, for example, how to achieve the highest possible recorvery of hydrocarbons.

Legislation adopted in the Soviet Union and its Federal Republics as well as the new Constitution emphasize the need for a careful approach toward the development of natural resources and their conservation for future generations.

The forecasted oil and gas reserves constitute only 6 and 5%, respectively, of the total reserves of fossil fuels on our planet (not including nuclear fuel). At the same time, oil and gas comprise 60 to 70% of the overall fuel and energy balance of most highly developed countries.

Production recovery and consumption of oil and gas in the world have increased by a factor of four over the past 20 years, and considering yields attained to date, recovery rate of oil and gas will apparently peak before the end of the century.

It is important to keep in mind that natural gas reserves are nonrenewable and finite, especially considering the rapidly increasing requirements for raw materials and energy for present and future generations.

According to data in this country and abroad, the average gas recovery-rate of completed fields and fields under development is estimated to be of the order of 0.85. In some fields (Maikop, Leningrad, Korobkov, and others), however, the gas recovery rate is significantly lower (0.6–0.8). The condensate recovery-rate of the reservoirs is much lower and in most cases the ultimate rate of recovery of gas-condensate fields operated without pressure maintenance is only 40–60%.

Thus, traditional technology for developing gas and gas-condensate fields in this country does not provide sufficiently high recovery-rates of gas and condensates to meet present demands. Traditional technology is particularly unsuited to the development of the new types of fields that have recently begun to be exploited.

The new types of fields of importance to the gas production industry include gas-hydrate fields. The gas-hydrate contribution to the overall production of gas in the USSR is still negligible, especially since after a ten-year period the Messoyakh gas-hydrate field is the only such field in commercial development. The prospects for future discovery of gas-hydrate fields are, however, very good.

In the Messoyakh field, part of the gas reserves are in a free state and part are in the form of hydrates. When gas is withdrawn from the gas-saturated zone, the reservoir pressure descreases. This leads to phase transformations in the hydrate-saturated zone. Decomposition of the hydrates is accompanied by the transformation of gas bound up in the hydrate into a gaseous phase and by flow into the gas-saturated zone. The reservoir temperature decreases with decomposition of the hydrates [1]. Preliminary research indicates that to attain a high gas recovery-rate from gas-hydrate fields, the development process will require artificial treatment of the reservoirs. Without such action, the gas recovery-rate of some gas-hydrate fields will not exceed 50%. For this reason, the process of decomposition of hydrates in a porous medium must be studied more thoroughly, and research to increase the gas recovery-rate of gas-hydrate fields is urgently needed.

Fields with anomalously high reservoir pressure (AHRP) are new to the gas industry. Research shows that gas fields with AHRP must be developed by external action on the processes taking place in the reservoir [2].

The "sealed" gas fields with AHRP, which were recently discovered in central Asia, deserve the greatest attention. Evidently, such fields will also include the fields confined to the Caspian Basin. One of the main features of the development of gas fields with AHRP will be increased deformation of the productive reservoir. This is because under high intrapore pressures part of the formation pressure is transmitted to the reservoir framework. Fields with AHRP often consist of carbonates and fractured-porous formations. Therefore, under traditional technology, as the reservoir (intrapore) pressure decreases, fissures will close and the reserves in the rock matrix will be gradually excluded from the draining process.

Analysis of results obtained from exploratory wells in the Zevard field shows that as the depression on the reservoir increases to a certain limit, the gas production rate increases. With further increase of the depression on the reservoir, the production rate does not change and then begins to decrease. Such production curves are interpreted based on the results in [3]. The corresponding interpretation of the production curves allowed forecasting the change in the well production characteristics as a function of the reservoir pressure. These results indicated a significant decrease in the gas recovery indicators as a result of reservoir deformation. Gas withdrawal from the Zevard deposit using traditional technology becomes profitable only with an ultimate gas recovery-rate of about 45%. It is clear that such efficiency is not acceptable.

In order to avoid reservoir deformation, a method has been proposed for developing natural gas fields with AHRP by maintaining reservoir pressure by means of water flooding. In contrast to known methods, this method for developing gas fields with AHRP permits increasing the ultimate gas recovery by maintaining optimum reservoir pressure by injecting water. Ultimate gas recovery-rates are increased due to the fact that pressure maintenance in the gas reservoir begins when the reservoir pressure decreases to the optimum value, i.e., the value corresponding to maximum economic utility from withdrawal of all hydrocarbon-components.

The proposed method is realized as follows for the Zevard field.

1. Development of the depleted field begins with an inventory of operational wells, whose number is determined by the planned recovery of gas.

2. When the reservoir pressure decreases to the optimum value indicated, maintenance of reservoir pressure by water injection begins. Thus, for the Zevard field, the optimum maintainance pressure equals 35.2 MPa with initial reservoir pressure equal to 50.2 MPa.

3. The water injection wells are perforated in the initial gas-water contact zone.

4. The volume of water injected equals the volume of gas produced from the reservoir (at the flooding pressure) since reservoir pressure will not decrease starting from the moment water is injected, naturally, the reservoir will not deform and the capacitative and flow parameters of the reservoir will not change.

5. Gas is withdrawn from operational wells spaced in the dome region of the gas reservoir. In so doing, the wells are opened by perforation of the order of 70 m in the gas-saturated thickness of the top part of the productive section.

6. As pressure is maintained in the flooded part of the reservoir, microscopic trapped volumes of gas are formed. The gas is withdrawn after water injection ceases and the pressure in the flooded zone of the reservoir decreases by 25–35% of the injection pressure.

Therefore, maintenance of the reservoir pressure for quite a definite time period is essentially important. As a result, it is possible to avoid decreasing the capacitive and flow parameters of the reservoir during the optimum period and avoid degrading the productive characteristics of the operational wells and it is possible to decrease their number. A considerable amount of gas is produced while reservoir pressure is maintained.

Withdrawal of gas trapped in the flooded zone takes place with further development of the reservoir under depletion drive. This possibility was proved earlier in laboratory experiments carried on the behavior of trapped gas with decreasing pressure in the model (element) of the reservoir [4].

The economic efficiency of variants utilizing water flooding, referred to the first year of the development, changes from 1278 to 1430 million rubles depending on well construction, tubing diameter, duration of the development, level of the pressure maintained, duration of water injection and volume of water injected, intake capacity and inventory of injection wells, etc. The largest economic efficiency of the variants examined (1421 million rubles) is obtained for the variant in which pressure is maintained at 35.2 MPa, the tubing diameter is 73 mm, the duration of injection is 15 years, and the volume of water injected is 471 million cubic meters.

The variant of development with pressure maintenance gives an increase in the economic efficiency referred to the first year by 65 million rubles. Due to enhanced rate of recovery of gas and condensate, an additional 64 billion cubic meters of gas and 3020 thousand tons of condensate are recovered. This means that the ultimate gas and condensate recovery-rates increase by 0.324 and 0.25, respectively.

In this variant, maintenance of reservoir pressure begins at 35.2 MPa in the 13th year, and ends in the 28th year of development. The volume of water injected per year is 26.5 million cubic meters and the daily volume is 71.5 thousand cubic meters. The required number of injection wells is 113 with an intake capacity of 595 m^3/day. The average cost of the gas recovered is 5.8 rubles/1000 m^3 and that of the condensate is 16.1 rubles/ton; the cost of the water injected is 11.5 rubles/1000 m^3.

Calculations show that even if the Zevard field were purely a gas field, it would still be advisable to maintain reservoir pressure by injecting water. It is for the first time that the problem of injecting water arises for a purely gas field.

Thus, in gas fields with AHRP, it is advisable to maintain reservoir pressure by water injection on a large scale. At the same time, the reliability of forecasts and conclusions as to the perspectives of developing the Zevard field and other fields with AHRP could be a matter of controversy. This is because no deposit with AHRP has yet been fully developed with the traditional or recommended technology. In this connection, the unique Pamuk field discovered in Uzbekistan is of great interest. It may be said that in this field, nature itself has arranged an experiment on changing the flow properties of the reservoir [5].

Three reservoirs have been explored in the Pamuk field: a gas-condensate reservoir (on the southern dome of the rise), a gas-condensate reservoir with an oil edge (central dome), and an oil reservoir with a gas cap on the northern dome. The southern dome reservoir is characterized by AHRP, and the initial reservoir pressure is 49.8 MPa. In the central and northern domes, the reservoir pressure is 29.9 MPa and corresponds to the hydrostatic pressure. The productive reservoirs in the Pamuk field consist of upper Jurassic carbonate rocks. The flow and capacitive characteristics of the cores of all three domes, determined at amospheric pressure, are practically identical.

The wells in the southern dome are characterized by very high productivity. For example, well-19 (a 6 meter interval was opened), gas production-rate of 2592 thousand cubic meters per day was obtained with a reservoir depression of 3.5 MPa.

The central dome is characterized by low flow parameters. The maximum gas production-rate obtained during well tests (well-11, the opened interval equals to 8 m) was 319 thousand cubic meters daily with a reservoir depression of 9.53 MPa.

It follows that during exploitation, of the deposit in the southern dome, when the reservoir pressure in the dome reaches 29.9 MPa, the wells must have production characteristic equal or close to the production characteristic of the wells in the central dome. Therefore, it is now possible to estimate the degree to which reservoir deformation affects the flow characteristics of reservoirs in the southern dome during exploitation. For this, the flow coefficients of an average well in the dome were calculated. Their comparison shows that when the reservoir pressure in the southern dome decreased from 49.8 to 29.9 MPa, i.e., to 0.6 Ph, the flow coefficient A increased by a factor of 56.2 and B by a factor of 1491. For the Zevard field, under comparable conditions ($\tilde{P}/Rn = 0.6$), A increases by a factor of 3.3 and B by a factor of 29.2. Thus the predicted flow coefficients used in planning the development of the Zevard field are not too high. The conclusion that water must be injected into gas fields with AHRP thus remains valid. As already noted, forced development of gas fields with AHRP is highly efficient economically.

Fields such as the Astrakhan field are encountered by the gas industry in the USSR for the first time. This field can be termed a sulfur-gas condensate field and is characterized by considerable reserves of gas, condensate, and sulfur, anomalously high reservoir pressures and temperature, and a high content of acidic components. The estimated value of the initial condensate and hydrogen sulfide reserves greatly exceeds the estimated value of the dry gas reserves in terms of money.

The study of the component composition of gases of more than 200 fields in the Soviet Union and the procedure based on this study for identification of a field type classifies the Astrakhan field as a gas-condensate field with an oil bank of a considerable size. The uniqueness of the Astrakhan field predetermines the necessity of a non-traditional approach to the basic principles of its planning and development.

The method for developing the Astrakhan field, the production volume, and the gas supplied to the national economy must be interrelated. The method in question must also satisfy the requirements of maximum recovery of sulfur and liquid hydrocarbons components, and compensate partially for the deficit of condensate and oil. Protection of the environment should be taken into consideration. This, to a large extent, predetermines the necessity of developing the Astrakhan field by pressure maintenance.

For the Astrakhan field, it is necessary to examine variants with total and partial cycling. In this case, annual gas production levels can be high enough for production of sulfur for export as well. At the same time, it is possible to produce in this field many tons of condensate (and evidently, oil as well). Based on calculations of the process of differential condensation of the reservoir mixture, we established that the condensate recovery factor equals 0.57. According to the regression dependences, built from the data on condensate recovery-rates from more than 120 gas-condensate fields in the USSR and other countries, the recovery factor is even lower, 0.33. Therefore, gas should be recycled in order to increase the condensate recovery rate.

The system in which reservoir pressure is maintained by injecting water may also be useful for the Astrakhan field. It is due to the anomalous reservoir pressure and the possibility for deformation of the producing reservoir. Water must also be injected if the presence of an oil bank is confirmed.

In studying the variants of the cyclinc process and water flooding of the reservoir, the high CO_2 content of the gas raises the question of pumping CO_2 back into the reservoir. If the reservoir characteristics have low values, then significant reserves of gas and accompanying components could remain in the zones of the reservoir located within the flooded regions. Pumping CO_2 into the corresponding injection wells displaces the reservoir gas toward the bottom of the operational wells and increases the gas and component recovery-rate of the reservoir.

Because of the unique composition and properties of the Astrakhan deposit, in formulating a strategy for its industrial exploitation, the development of the reservoir cannot be separated from the system for refining the well production. A comprehensive systems approach is required.

This means, first of all, that it is necessary to formulate a goal and to determine clear criteria for choosing both the system for developing the field and the technology for processing the well production.

To choose a method for refining the production of the field, it is necessary to have highly reliable data on the effieciency of different methods for the specific conditions of the Astrakhan field. For rapid and efficient bringing into operation the Astrakhan field, it is useful to equip several wells in the main gas-bearing areas with plants having alternative methods for refining the well production. Pilot operation of wells following special programs will enable a choice of optimum technology for refining of reservoir production.

The need to change the traditional approach to development of depleted natural gas fields is also dictated by the forthcoming bringing into operation of the Karachaganak gas-condensate field. This deposit has a unique gas-condensate characteristic. The specific condensate content of the reservoir gas exceeds by more than a factor of two its content in the gas from the Vuktyl' field. The composition of the reservoir fluid indicates the possible presence of an oil bank in the field.

For the Vuktyl' field, in the depletion regime a condensate recovery-rate of about 0.33–0.35 will be obtained, i.e., lower than the average oil recovery-rate in the USSR. It is well known that condensate recovery-rate decreases as the specific condensate content with high specific condensate content, a steep isotherm of reservoir condensate losses is also characteristic. This indicates that a large part of the condensate precipitates into the formation immediately when the reservoir pressure begins to decrease. This confirms that the Karachaganak field cannot be developed in the pressure depletion regime even during experimental pilot operation (EPO).

The absence in this country of high-pressure compressor units with anticorrosion construction does not now permit the most efficiency cycling process as a development method for the Karachaganak field.

The most realistic method now for the Karachaganak field (at least during the EPO period) is maintenance of the reservoir pressure by injecting water. This deposit is unique in that one cubic meter of pore space at the initial reservoir pressure contains 300–400 kg of liquid hydrocarbons (stable condensate). If we take into account that one ton of condensate is equivalent to 3–5 tons of oils, then the equivalent content of liquid hydrocarbons per cubic meter of pore space in the Karachaganak field is several times greater than the oil content per cubic meter of pore space of any oil deposit. No one doubts the necessity of water flooding of oil fields, and this method is widely used in oil recovery practice. Reservoir pressure maintenance in the Karachaganak field by water flooding is all the more justified.

Establishing the presence and estimating the practical value of the oil edge in this field will require considerable wells and time. The corresponding investigations will apparently be completed later the EO period. Development to depletion will decrease the reservoir pressure, displace the oil edge into the gas-bearing region, and, therefore, decrease the oil recovery-rate. This circumstance also raises the question of water flooding the Karachaganak field during the EO period as well.

The proposed injection of water and corresponding monitoring of the flooding process will give data on the reservoir-mass type of the reservoir and will determine the time and interval encroachment of water to the observational wells. As a result, it will be possible to estimate the capacitive and flow characteristics of separate layers. This will permit to evaluate more precisely the gas and condensate reserves according to the volume method and estimating the intensity of the water-drive in the future. Thus the proposed method for gas-hydrodynamic exploration will yield important information on the structure of the field which is necessary for final justification of a method for developing the Karachaganak field.

It will be necessary to inject about 16000 cubic meters of water daily into the field to recover 3×10^9 m^3 reservoir gas annually. Injection of this quantity of water will provide an annual production of about 2.5 million tons of condensate or, referring to oil, approximately 10 million tons of oil.

It is expedient to maintain reservoir pressure so that the barrier water flooding variant is realized. In the absence of any sign of gas-oil contact in the first injection wells, it is permissible for water "bubbles" to form near the wells. The creation of such water saturated zones facilitates solving the problems of gas–hydrodynamic exploration of the reservoir, as indicated above, and estimating its accumulation properties.

Laboratory and theoretical studies of the development of gas fields with inflow of edge or bottom water have been carried out at the I. M. Gubkin Institute of Petrochemical and Gas Industry in Moscow since the end of the 1960's [1, 6, 7].

Analytical studies show that it is possible to affect the dynamics of motion of the gas–water interface by controlled interdistribution of gas recovery from separate wells of the fields. It is possible to increase the gas recovery-rate under the water-drive.

Here, the creation and use of a self-adjusting, comprehensive geological and mathematical simulation of the deposit was effective. This approach was first realized for the Medvezh'e field [8]. An adequate geological and mathematical simulation of the field is created based on actual data obtained from the development of the field. Such a model yields a reliable forecast of the following development indicators:

– time of appearance of bottom water in the production of some operational wells;

– reservoir pressure and temperature in the, pressure and temperature at well bottom and heads, and at the inlet and outlet of the corresponding plants for complex refining of the gas (PCRG);

– pressure at the inlet to the head compressor station (HCS), time at which the HCS is placed in operation, and the power of the HCS as a function of time;

– time at which booster compressor stations (BCS) are placed in operation and the power of BCS at separate PCRG as a function of time;

– completion time of development of the field and final gas recovery-rate;

– cost of the gas production, capital investment, and operational expenditures on deposit development and field facilities as a function of time;

– reduced expenditures and the economic efficiency for the development variant under study.

The creation of a self-adjusting comprehensive, geological–mathematical simulation of the Medvezh'e field permitted mathematical computer experiments on optimizing the indices for further development of the field. Simulation to estimate the consequences of technological decisions enabled validation of the optimum strategy for further development of the Medvezh'e field.

Redistribution of gas recovery from the field over separate PCRG, increasing the period of time with constant gas production by three years, and drilling an additional twenty operational wells decreases the inventory wells with water encroachment, shortens (compared to the development plan) the period of time for developing the field by fifteen years, increases the final gas recovery-rate by 2.5%, and increases the economic efficiency referred to the first year by 240 million rubles.

Thus the creation and widespread use of self-adjusting comprehensive, geological–mathematical models of the largest fields in the USSR in combination with simulation opens up possibilities of greatly increasing the effectiveness of all gas recovery indices.

Studies of the problems of controlled development show that the system of gas bearing reservoir–water-drive basin is very conservative so that great effort must be made to approach the desired results from the water-drive. Because of the large difference between the viscosity of gas and water it is difficult to affect essentially the motion of the gas–water interface due to the creation of pressure gradients in the gas-bearing region.

Therefore, it is impossible to achieve the most vital results from the water-drive because of the changes in the technology regimes of producing well. For this reason, an effort was made to create active methods for treatment the processes in the gas reservoir accompanied by water encroachment [4].

The method proposed for controlling development under the water-drive is based on results of laboratory experiments [6]. In these experiments, the behavior of the trapped gas in the model with water encroachment of the reservoir was investigated. The experimental results show the following.

When pressure in the model in question of the reservoir decreases, bubbles of trapped gas begin to expand. The coefficient of residual gas saturation increases, while the phase permeability to water decreases correspondingly.

After the pressure in the model is decreased by 25–35% of the flooding pressure, the gas bubbles begin to coalesce forming gas-saturated "channels" for the trapped gas to reach the outlet of the reservoir model. At the same time, the coefficient of residual gas saturation and phase permeability to reservoir water are stabilized and practically do not change with further decrease in pressure in the model.

Laboratory experiments indicate that when the pressure Z decreases to atmospheric pressure in the water-flooded model, the gas recovery-rate relative to the trapped gas approaches unity. In reservoir models with a pressure decrease of 25–35% (for different experiments) the water phase permeability decreases by a factor of ten and higher.

The above mechanism of behavior of the trapped gas and the corresponding quantitative characteristics allows to propose a new method for development of the gas reservoir under water drive [4]. The essence of the method under consideration that it proposes the necessity to operate water flooded wells, for the purpose to recover the underground water.

When recovering reservoir water from a flooded well (for example, by a gas-lift or mechanical method), further motion of the corresponding water fingering into the formation is prevented or slowed down. Because the compressibility of water and gas differ by several orders of magnitude, the reservoir pressure decreases quite rapidly in the vicinity of the well.

This leads to expansion of trapped gas bubbles. After the bottom pressure decreases by 25–35% of the reservoir pressure, the trapped gas acquires mobility and begins to enter the well. The expansion of gas bubbles decreases the phase permeability to water in a large area around the well. Therefore, zones with worse flow characteristics (which also prevent the motion of water into the reservoir) form around each operational water-flooded well.

The practicability and efficiency of active treatment of a water-bearing reservoir are already indicated today by indirect data on the gas-condensate fields in the Krasnodarsk region. The water-drive was developed actively and selectively in these fields. As a result, not only micro-, but also macro-volumes of trapped gas remained in flooded zones of the reservoir. In recent years, low reservoir pressures were attained in a number of fields. For this reason, the macro- and micro-trapped gas is transformed into blocks of gas which are in a moving state in the flooded zones of the reservoir. As a result, some of the previously flooded wells in the Maikop, and especially Berezan, fields are becoming operational [9].

Actual data on the fields in Krasnodarsk region support the idea of active treatment of the water-drive basin, although in these fields the treatment of the drowning wells was passive. They started to produce gas (with water) as a result of a general, natural decrease in reservoir pressure and, therefore, in the zones of the reservoir with water encroachment as well. The effectiveness of forcing the drowning zone of the reservoir is also indicated by data from recent operation of a number of wells in the Bitkov field that are being water encroached [10].

Thus the following conclusions were drawn from experimental data on wells in the Orenburg field. Wells, whose production contains water, should not be shut down. Gas production rates of wells showing water should not decrease. This does not decrease the water content of well production. Experience shows that decreasing the production rate of a well too soon often leads to self-sealing of well by reservoir water [11].

It is difficult to estimate now, based on field data, the efficiency of introducing into operation the drowning wells Nos. 508, 509, 503, 229 and others in the Orenburg field from the point of view of their influence on the final gas recovery-rate of the reservoir. This is related to a number of reasons. First, the scale of the field experiment is not yet very large. Second, estimates based on, for example, a geological field analysis or gas-hydrodynamic calculations are difficult to make in the absence of reliable data on the capacitive and flow properties of the reservoir in the region of PCRG-8, etc.

In this connection, it became necessary to conduct mathematical computer experiments to investigate the effect, on the magnitude of the final gas recovery-rate of the production of formation water from wells [12]. The theory of two-phase flow was used for this purpose.

The calculations were performed on a two-dimensional model of a rectangular reservoir in a plane. The reservoir is piece-wise nonuniform with respect to the formation properties. Prior to water encroachment to the wells, the reservoir is developed with a constant gas recovery rate equal to 5% of the reserves per year. The production rates of all 18 wells are identical and equal 40 000 m^3/day. The production rate of edge water is given as a function of time $q_w = q_w(t)$ along the perimeter of the reservoir.

For this two-dimensional model, two variants of the calculation were performed. In variant I, the popular ideology of developing gas fields with water-drive is used. This means that after drowning any well, this particular well is removed from the inventory of operating wells. The development indices are calculated to the time that the last operational well is flooded.

In variant II the development of the gas reservoir in question is modeled following the idea of active water-drive. After water-encroachment into each of the operating wells, they are not shut down, while water is produced from them at an identical, constant rate equal to 100 m^3/day. In the corresponding forecasts, the limitation for completion of development (recovery of gas, water and gas) is the attainment of reservoir pressure equal to 1 MPa in the vicinity of the wells.

The computational results for these variants of development show that after 17.3 years of development, in variant I only two wells are operating in the zone of the reservoir without water encroachment, while in variant II, five wells are operating. The large gas-bearing zone in variant II is explained by the fact that part of the water entering the reservoir is produced through the wells.

The high recovery of gas from the wells without water encroachment in variant II accounts for the fact that the gas for the date examined is 82%, while in variant I it is 76.4%. This difference is also related to the fact that in variant II trapped gas is recovered when water is produced from the drowning wells.

Before drowning the wells, as noted above, in both variants the rate of development of the reservoir is identical. After water encroachment into the wells begins, there is a sharp drop in gas production in variant I. Development is terminated in variant I after 18.6 years.

Due to the operation of drowning wells, the duration of development increases by 8.6 years in variant II. It is interesting to note that gas was recovered at the expense of operating the drowning wells the two last years.

Predictive indices of development for variants I and II indicate that the total amount of gas recovered in variant II greatly exceeds the cumulative recovery of gas in variant I. The ultimate gas recovery-rate is 77% in variant I and 97% in variant II.

During the period that the water-drive is active, about 30% of the water entering the reservoir is withdrawn in variant II. The value of this is that the ultimate gas recovery-rate of the reservoir increases by 20%. In addition, recovery of gas only with operation of drowning wells increases the gas recovery-rate by 9%. Due to a lower rate of decrease in the gas-bearing area, an increase in the inventory of operating wells, and an increase in the gas recovered over the period free from water encroachment, the ultimate gas recovery-rate increased by 11% in the second variant as compared with the first one.

In connection with water encroachment into wells and the reduction in gas recovery, the average reservoir pressure in variant I begins to increase after 14 years of development. On the other hand, in variant II the average reservoir pressure continuously decreases as a result of the partial production of water.

The operational characteristics of drowning wells are definite. The water saturation coefficient of the reservoir increases during the water-encroachment process at the well location. As a result, the gas production rate of the well decreases. After water production begins, the gas production rate of the well somewhat increases. Further decrease in the gas production rate is due to the decrease in the reservoir pressure near the examined well.

The results of the investigations presented permit estimating the effect of forcing the water-drive on the gas recovery-rate in the case of a single-reservoir formation. The studies completed support the forcing action from the point of view of maintaining the inventory of operational wells, and, most important, maximizing the gas recovery-rate of the reservoir.

The Orenburg field belongs to a category of fields that is unique with respect to the gas and condensate reserves and is distinguished by the complexity of geologic structure and presence of bottom water. The development of this field began according to the development plan for the proposed gas regime. To a certain extent it was assumed that the oil edge present would serve as a barrier to water inflow (in analogy to the Sovkhoz field). Actual data on the development of the Orenburg field indicate the presence of a selective and quite active water-drive.

The first signs of water-drive arose practically at the very beginning of the development of the Orenburg gas–condensate field (OGCF). As a result, reservoir water began to appear in the production of the operating wells. Such active, selective appearance of the water-drive has not yet been encountered in any known field. The presence of a high gas-bearing horizon, considerable gas reserves, an oil edge, and an anomalously high reservoir pressure should, it would seem, serve as factors preventing early motion of reservoir waters into the gas formation.

The present development of the Orenburg field can be described as follows. The gas enters the wells along highly permeable thin intercalations. From the low permeability reservoirs, the gas flows into these highly permeable intercalations due to the considerable size of the contact surface. Contribution of the low permeability reservoirs directly to the wells is apparently of secondary significance.

The mechanism for decreasing the gas and condensate recovery-rates of the reservoir in reference to OGCF is described as follows. Water, entering the reservoir along highly permeable intercalations, decreases the contact surface area of the high- and low-permeability reservoirs. As a result, the production rate of gas flowing from the low-permeability reservoirs into the high-permeability reservoirs decreases with time. In the limit, if the reservoir water crosses all high-permeability intercalations, then the gas reserves in the low-permeability reservoirs will be partially or completely cut off from draining. Drilling a large number of additional exploitation wells to recover gas from low-permeability reservoirs is a very inefficient process, requiring enormous capital investment and large operating expenses. Field experiments to bring the drowning wells into operation were begun in this connection [11].

We shall now examine the results of gas-hydrodynamic calculations, which include structural characteristics of the productive reservoir of OGCF. OGCF is represented as a layered reservoir where low-permeability differentials alternate with high-permeability intercalations. A working profile model of an element of the OGCF reservoir was chosen for this development scheme.

The inflow of water is simulated at an average rate of 138 m^3/day through the right butt-end of the reservoir element (for 1 m of reservoir width). Gas is withdrawn through the left butt-end at a rate of 15×10^3 m^3/day, which corresponds approximately to the present rate at which the OGCF is being drained. The permeability of the high-permeability intercalation is 10^{-12} m^2 and that of the low-permeability intercalation is 10^{-15} m^2, the length of the reservoir element is 2000 m, and the overall thickness is 125 m.

The mathematical experiments were performed on a computer in two variants.

Variant I is a model of a passive system for developing the chosen element of the reservoir. This means that gas is withdrawn only through the operating gallery (left butt-end). The model parameters are such that the gas in the gallery flows primarily along a highly permeable intercalation. Gas flows into the high-permeability intercalations from the low-permeability reservoir due to the pressure difference. Reservoir water enters into the high-permeability intercalation and, in small quantities, into the low-permeability intercalation.

Forecasts are continued until the reservoir water reaches the operating gallery (along the high-permeability intercalation). The calculations show that at this time, 47% of the starting gas reserves will be withdrawn from the reservoir element. Therefore, the final gas recovery-rate in variant I equals 47%.

Variant II is a model of the forcing approach to the development of OGCF. In contrast with variant I, an allowance is made for partial withdrawal of water entering the reservoir element. Water is withdrawn from the gallery at the center of the reservoir model. Water is withdrawn at a rate equal to 10% of the average rate of extraction of water entering the reservoir element. The forecasts are terminated when water breaks through into the operating gallery or a pressure of 1 MPa is attained.

In variant II, termination of gas extraction was predetermined by the limitation on the bottom pressure of the exploitation gallery. In this variant, the gas recovery-rate constituted 56%.

Therefore, withdrawal of 10% of the water entering the reservoir provides a 9% increment to the final gas recovery-rate. Of the 9% increment, 1.6% is for gas recovered together with water from the unloading (drowning) gallery. The remaining additional recovery of gas comes from the gas–water contact in the highly permeable interacalations, moving more slowly after water production begins. For this reason, more gas flows from the low-permeability reservoir into the high-permeability intercalation.

The expediency of active development of OGCF lies in the smaller inventory of wells required for gas recovery. This is because the operation of drowning wells and wells that are being encroached will be maintained by a higher number of undrowning wells, and an additional amount of gas will be recovered with the production of water. On the other hand, the main argument in favor of forcing is that that forcing predetermines the considerable increase in final gas recovery (component recovery) of the OGCF reservoir. It is clear that when all of the water entering the reservoir is withdrawn, the gas recovery-rate will approach 100%.

A natural question arises here: what volume of water withdrawn from the OGCF can be expected today and in the future? The following investigations were performed with this question in mind.

The OGCF was approximated by a layered reservoir. Due to the inadequate basic information, each of the three operating objects of the OGCF was represented as two low-permeability reservoirs with one high-permeability intercalation between them. The history of such a multilayer model of the OGCF was reproduced. The actual annual gas recovery by field was given up to January 1, 1980. Corresponding gas recovery rates were chosen for each of the operational objects from VUNIPIGAZa data.

Using the actual average reservoir pressures over the operational objects and data on the percentage of the inventory of water-drowning wells (gas-bearing areas with water encroachment in the high permeable intercalations), the parameters of the layered OGCF model were refined. Using these refined parameters, calculations were made estimating the amount of water entering into each operating object up to January 1, 1980 as well as projected into the future.

According to the 1980 level, the average daily flow rate of water into the OGCF is about 9000 m^3. It is hardly possible to withdraw all this water, since there is no intended drilling of special relief wells. We are concerned with withdrawing water from wells that have already been and are being encroached by water. Then, apparently, the daily produced volumes would not be greater and would not exceed 1–2 thousand cubic meters, but the production of water is a necessity. At the 1985 level, about 14 000 m^3 of water per day will enter the OGCF. This means that the increments to the production of withdrawal water will also be insignificant.

Thus the data presented indicate that forcing of processes occurring in gas fields is a matter of time. Widespread adoption of forced development is highly efficient economically.

References

1. Korotaev, Yu. P., and S. N. Zakirov (1981). *Teoriya i proektirovanie razrabotki gazovykh i gazokondensatnykh mestorozhdenii (Theory and Planning of the Development of Gas and Gas-Condensate Fields)*, Izd. Nedra.
2. Zakirov, S. N., Yu. P. Korotaev, E. I. Petrenko, M. M. Dzhalilov, and A. F. Samoilova (1979). *"Planning the development of gas fields with abnormally high reservoir pressure", Obzornaya informatsiya*, Izd. VNIIÉGAZprom, No. 7.
3. Korotaev, Yu. P., L. G. Geroev, S. N. Zakirov, and G. A. Sherbakov (1979). *Fil'tratsiya gazov v treshchinovatykh kollektorakh (Permeation of Gas in Fractured Reservoirs)*, Nedra.

4. Zakirov, S. N., Yu. P. Korotaev, R. M. Kondrat, V. N. Turnier, and O. I. Shmyglya (1976). *Teoriya vodonapornogo rezhima gazovykh mestorozhdenii (Theory of Water-Drive in Gas Fields)*, Nedra.
5. Zakirov, S. N., and E. I. Petrenko (1980). *"Effect of deformation on reservoir permeability" Geologiya i razvedka gazovykh i gazokondensatnykh mestorozhdenii (Geology and Exploration of Gas and Gas-Condensate Fields)*, VNIIGAZprom (1980), No. 2, pp. 8–14.
6. Trebin, F. A., S. N. Zakirov, R. M. Kondrat, and N. A. Manomenova (1970). *Issledovanie osobennostei proyavleniya vodonapornogo rezhima pri razrabotke gazovykh mestorozhdenii (Investigation of the Appearance of Water-Drive in Development of Gas Fields)*, Izd. VNIIÉGAZprom (1970).
7. Zakirov, S. N., G. A. Zotov, Yu. P. Korotaev, G. D. Margulov, A. N. Timashev, and V. N. Turnier (1972). *Voprosy razmeshcheniya skvazhin i analiza razrabotki na élektricheskikh modelyakh (Problems of Well Placement and Analysis of Development Using Electrical Models)*, Izd. VNIIÉGAZprom.
8. Korotaev, Yu. P., S. N. Zakirov, S. V. Kolbikov, A. I. Ponomarev, and R. G. Shagiev (1980). *Sostoyanie i perspektivy razrabotki mestorozhdeniya Medvezh'e (Status and Prospects of Development of the Medvezh'e Field)*, Izd. VNIIÉGAZprom (1980).
9. Ratushnyak, N. S. (1979). *"Mobility of trapped gas in the water-flooded zone"*, Gazovaya promyshlennost, *No.* 7 (1979).
10. Kondrat, R. M., I. N. Petrashch, T. L. Levitskii, N. S. Shvadchak, D. Yu. Demyanchuk, and M. V. Biletskii (1980). *"Experiment on introducing the gas-lift method for exploitation of water-flooded wells in the Bitkov field" Razrabotka i ékspluatatsiya gazovykh i gazokondensatnykh mestorozhdnii (Development and Exploitation of Gas and Gas-Condesate Fields)*, Izd. VNIIÉGAZprom, No. 9, pp. 30–34.
11. Vyakhirev, R. I., Yu. A. Dashkov, I. Yu. Zaitsev, S. N. Zakirov, Yu. P. Korotaev, B. E. Somov, Yu. V. Uchastkin, and E. E. Frolov (1979). *"Forcing of water-drive" ÉI Geologiya, burenie i razrabotka gazovykh mestorozhdenii (ÉI Geology, Drilling, and Development of Gas Fields)*, Izd. VNIIÉGAZprom, No. 20.

Sov. Tech. Rev. A Energy Reviews, Vol. 2, 1985, pp. 173–287
0275–7893/85/002–173 $30.00/0

TECHNICAL PROGRESS IN CONSTRUCTION OF MAIN GAS PIPELINES FOR THE UNIFIED GAS SUPPLY SYSTEM OF THE USSR

O. M. IVANTSOV

Minneftegazstroi, Moscow, Zhitnaya No. 14

1. Introduction

The Soviet Union is currently engaged in large-scale programs involving the construction of main pipelines. Ten to fifteen thousand kilometres of mainlines are being laid annually. This constitutes 20 to 25% of total pipeline construction on all continents.

Pipeline construction in the USSR has now crossed the 100 year mark in its history. The first oil pipelines, which were about 10 km long and had a diameter of three inches, were constructed in Russia in 1878 near the city of Baku. The first main gas pipelines were constructed during World War II: Elshanka–Saratov, Kurdyum–Saratov, and Pokhvistinevo–Kuĭbyshev. It should be noted that during the difficult years of World War II, the pipeline network in the Soviet Union was extended by more than 1500 km.

The first long-distance gas pipeline from Saratov to Moscow, which is 800 km long and has a diameter of 325 mm, was put into operation in 1946. The construction of the Stavropol'–Moscow and Serpukhov–Leningrad gas pipelines played a large role in the development of Soviet pipeline transportation. Large-diameter pipes and special construction equipment and procedures were used here for the first time. By the mid-1960s, more than 7200 kilometers of pipeline were in operation in the Soviet Union.

At the present time, the total length of mainlines has reached 2200 km, of which gas pipelines constitute 1400 km. One half of the total length of mainlines was laid during the last decade. Now, predominantly large-diameter and high-pressure gas lines are being laid. As a result, pipeline efficiency, referred to the base pipeline diameter and capacity in the 1970s, has quadrupled.

More than 12 000 km of the largest gas pipelines in the world, having a diameter of 1420 mm, working pressure of 7.5 MPa, and a carrying capacity of 32–36 billion cubic meters per year have been constructed.

The largest gas transportation systems have been constructed in the last few years: Urengoĭ–Punga–Ukhta–Torzhok, Urengoĭ–Punga–Nizhnyaya Tura–Perm'–Kazan'–Gorki–Moscow, Urengoĭ–Vyangapur–Chelyabinsk–Petrovsk–Novopskov, Urengoĭ–Gryazovets–Moscow district terminus, and Soyuz gas pipeline from Orenburg to the western border of the USSR (in cooperation with the member countries of Comecon).

The length of individual gas pipelines has now reached almost 3000 km. The program for constructing main gas pipelines, branches from the mainlines, as well as industrial pipelines is continuously growing.

As a result of the completion of a large-scale construction program in the 1970s, a new industry – the pipeline transportation industry – has been created. Today's national pipeline transportation system is distinguished by high efficiency, new engineering techniques and technology used in its construction, and unique mainline capacity.

The energy equivalent of the gas flow in a pipeline with a diameter of 1420 mm and working pressure 7.5 MPa is 15.2 million kW. Gas pipelines

Table I Characteristics of Gas Pipeline Transportation

Year	Length (1000 km)		Volume transported* (million tons)	Turnover* (billion ton-km)
	Nominal	Reduced		
1	2	3	4	5
1970	67.5	67.5	204.7	187.7
1975	98.7	145.8	305.0	377.3
1980	132.7	245.1	440.0	795.0

*Scaled to reference fuel.

of this diameter constitute 11% of the total effective length of pipelines, and more than 40% of all gas by volume and 50% by turnover is transported along them. If this transportation problem had been solved using pipelines with lower capacity, for example, a diameter of 1220 mm and a working pressure of 5.5 MPa, then it would have been necessary to construct twice as many pipelines.

The transportation work performed by main pipelines in 1980 amounted to 2.6 trillion ton-kilometers referred to TUT. With respect to turnover, pipeline transportation is now second to railroads.

Looking at the trend in the development of transportation in the Soviet Union over the last 20 years, it is evident that the relative contribution of railroads to the total turnover has decreased from 79.4% to 50.4%, while relative turnover of pipeline transportation has increased by a factor of 8 and it has surpassed the combined turnover of river, marine, and automobile transportation.

The forecast shows that by 1990 the relative contribution of pipeline transportation will increase by a factor of 2.3–2.6, the gas turnover will increase by a factor of 3.3–4.0, and the average transportation distance will increase by a factor of 1.4–1.5. Data on the development of gas pipeline transportation over the last decade are presented in Table I.

Due to the high rate of construction and an increase in the technological level of main gas pipelines, the gas supply to consumers has been centralized.

Separate gas fields, underground storage facilities, and main gas pipelines and their local subsystems are united into a single gas supply system, which has opened up the possibility of interregional maneuvering of resources and has increased the reliability of gas supply.

Western Siberia has become the main oil and gas producing region. In

1985, 330–370 billion cubic meters of gas and more than one-half of the planned production of oil in the USSR will be produced there in 1985.

To supply Siberian gas to the central regions of the country and for export, the construction of six large main gas pipelines with a diameter of 1420 mm pressurized at 7.5 MPa and a total length of about 22 000 km is planned in 1981–1985. The construction of gas pipelines pressurized at 10.0 MPa will begin at the end of the five-year period.

The environment and climate of Western Siberia are distinguished by their uniquely complex conditions for equipping gas fields, laying main pipelines, and constructing compressor stations.

Arctic and subarctic regions are covered by permafrost and abound with marshes. Forty-five per cent of the territory of Western Siberia is covered with impassable swamps. In the Tyumen region, there are more than 300 000 lakes and other reservoirs. For every 100 km of gas pipeline, in the Ob' River basin, there are 37 km of different types of swamps, 10 km of rivers, streams and lakes; in the forest-tundra zone, there are 53 km of marhses and flooded areas and 2 km of permafrost.

The carrying capacity of the soil over most of the territory of Western Siberia during the summer is less than 0.5 kg cm^{-2}, which makes it impassable for transport and wheeled and heavy tracked machinery. The soil temperature under the natural cover at a depth of 0.4 m is, on the average, depending on the region, negative for 152 days of the year. Construction occurs primarily during the winter months. In the Yamalo–Nenetskiĭ Autonomous District, 120–140 days of the year are favorable for pipeline work and in the Khanty–Mansiĭ District, there are 100–120 favorable days. On the average for a region, there are 14 days with a temperature of $-40°$ C and, in addition, work is prohibited in the open air with a wind force of 6 (about 14 m s^{-1}). There are 14 such days as well. In addition, work stops for 6–8 days of the year due to fog and precipitation. The total number of idle days, taking into account the factors indicated above, constitutes 15–20% of the winter work time. Thus the time available for pipeline work in Western Siberia is very limited and it is not much greater for constructing compressor stations and construction in gas fields.

The general problem is clear: to achieve a gas production level of one trillion cubic meters over the next 10 to 12 years. Fulfillment of this program is becoming an important factor in the development of the country in the 1980s.

The economic value of gas pipeline transportation systems will increase together with the contribution of gas production to the fuel-energy balance of the country. They will undergo qualitative changes and they will utilize new technology.

It is well known that in the USSR, there has been an extensive quest to increase the efficiency of long-distance gas transportation, and it has been proposed that pipelines with diameters up to 2.5 m be built and that gas be transported in cooled and liquified form. The most urgent problem is the construction of gas pipelines with a diameter of 1420 mm pressurized at 10.0–12.0 MPa.

The development of a gas transportation system with an annual capacity of one trillion cubic meters will change the structure of the fuel balance of the country, eliminate the fuel-energy deficit in the European part of the country, decrease the consumption of oil as an energy fuel, and increase exports. This is the first construction program with such a large scale and large economic impact.

To solve the problems posed within the short time periods planned, it is necessary to use a new, higher level technology and better construction organization and to create a better economic infrastructure which addresses the needs of this program. In the Soviet Union, all main pipelines are laid by the construction-assembly organizations of the Ministry of Construction for the Oil and Gas Industry (MCOGI).

Preparations are currently underway for construction of main pipelines using the "turn-key" method. The intention is to combine under a single management organization all stages of the investment process and to place the responsibility for research, design, setting aside the necessary land, engineering and economic preparation, construction and assembly work, providing all required equipment, and the startup-management work including delivery of the completed main gas pipelines in the hands of MCOGI.

This will increase the reliability and technical level of design work, ensure high construction quality and reliability of the systems created, and sharply decrease the time required to put gas pipelines into operation. Reduction of the time required to construct mainlines is a very important efficiency factor. Economists assert that if a 1420 mm gas pipeline is delivered one year ahead of schedule, then the saving amounts to more than one billion rubles. To achieve the highest efficiency in gas industry construction, in recent years, the management scheme for construction of gas mainlines has been changed.

The construction organizations are specialized not according to the type of work, but according to the stages of construction. Subunits, corresponding to the linear nature of the work, are mobile, and the economic productivity of the subdivisions is evaluated according to the finished product output. The quality control system has also been improved. Provisions have been made for further development of the machine building industry, construction organizations, and scientific-research and design-construction organizations.

Executing the largest program of modern times, namely, development of the Western Siberian oil-gas complex, requires new methods for organizing and managing all work in the region and, in particular, mainline construction.

Increasing the length of pipelines has not changed the ultimate problem of producing gas in this region and providing for ever increasing annual increments of gas.

Of course, the productivity of the operational gas pipelines will increase as new compressor stations are constructed and old ones are upgraded or as loops are constructed. However, this does not solve the problem of delivering the total annual increment of gas to consumers. In practice, there arises a problem which has never before appeared in the history of the gas industry: construction and putting into operation within one year a gas pipeline with a diameter of 1420 mm. And, if the entire annual increment of gas must be delivered to consumers, then the length of the pipelines laid from the Tyumen region will exceed 2500 km.

Therefore, the linear part of the pipeline and the compressor stations as well must be constructed as rapidly as possible, and this must be accomplished without using large amounts of labor. The problem is complicated by the seasonal nature of the work on long sections of the path.

In line construction, large mechanized columns (complexes), which perform all technological operations, have been very successful in practice. Here, a column (complex) includes all personnel and machines which are concentrated on a single section in order to extend the pipeline by a certain amount.

The criterion for choosing the optimum organization for line construction was the organizational-technological tension, indicating the ability of the system to conserve its technological unity in the face of organizational separation of the process, as well as the degree of technological specialization.

Technological specialization permits maintaining a high degree of readiness, permits making good use of special technical resources and specialists, and creates favorable conditions for broad assimilation of technological achievements.

The greatest gain, as indicated by an integral indicator, was achieved by using complex trusts, which include management of welding-insulation and road construction work and specialized subunits for excavation and preparatory engineering work on the route, to manage the work. The first complex trusts were organized within the Glavsibtruboprovodstroĭ and their work has completely confirmed the advantage of such a structure.

In recent years, large mechanized columns have been laying up to 5000–6000 kilometers of pipeline per year.

The short winter season, which is advantageous for line work in the North, made necessary a closer interaction of the production resources engaged in construction.

However, concentration of resources makes it less likely that a stable rate will be achieved. The largest fluctuations in insulation-laying operations occur in the middle belt of the country, whereas in the North excavation operations are subjected to the greatest variations in average daily rates.

Concentration of resources on a small part of the line is most advantageous under unified management, as provided for in large-scale mechanized columns, since it is easier to make optimum use of resources by arranging separate work functions so as to concentrate on the critical operations. Increasing the construction capacity of a column increases line construction efficiency because of lower expenditures on servicing the construction operation.

To increase the organizational-technological reliability of the work of such large subunits, which in the North includes about 600 laborers, engineers, and technicians and up to 310 machines and pieces of machinery, the optimal reserve of the basic technological and auxiliary machines is determined according to a special procedure. The use of this procedure increases the pipe-laying rate by more than 10%.

During the last three years, separate complexes of Glavsibtruboprovodstroĭ laid more than 100 km of 1420 mm pipe in the North during the winter under difficult working conditions.

Generalization of the experience and various studies have shown that the optimum solution is limited to a rate of about 1 km/day. When the output increases to 150–200 km, it is necessary to shift the complex to a period that is most favorable for performing line work, which greatly decreases the efficiency of its operation. The daily rate indicators mentioned above are not the limiting values. Separate columns on the Vyngapur–Chelyabinsk gas pipeline welded and insulated up to 2 km per day, and a rate of 2.5-3 km per day of finished pipeline was achieved in constructing other large-scale gas pipelines.

In practice, however, there are numerous examples when protracted scrubbing and testing operations held up completion of the pipeline for a long time. Hydraulic testing of gas pipelines has greatly decreased the testing time, for example, by a factor of 3 compared to tests using air, which saves about 2 million rubles per year.

The problem is to include pipeline scrubbing and testing in the main work complex. Such a system for laying large-diameter pipelines will decrease the construction time from one to two months per 1000 km. The complex includes separate mechanized columns (brigades), which perform technologically interrelated operations in parallel.

These subunits are linked in a complex dynamic manner and, taking into account the fact that the subunits contain expensive machinery, which when standing idle greatly decreases the economic efficiency of the construction operation, it becomes necessary to choose an optimum structure for a column that conforms to different pipeline parameters and specific construction conditions.

To solve such problems, a mathematical model of a large-scale mechanized complex was developed and the required algorithms were constructed.

The coupling between separate columns (brigades) characterizes the structure of the complex mechanization.

The basic requirements imposed on the organization of mechanized work are: continuity of construction, proportionality between all technological links, and smooth execution of the work.

Fulfillment of the basic principles of organization of line construction facilitates development of operational and emergency reserves of machinery. The size of the line stock has increased and its quality has improved. The active part of the basic production reserves (machine stock) has reached 1 billion rubles.

The system for operational-production planning and management in pipeline construction is being checked in practice. This system is based on finding a method for scheduling and operational management that permits improving the utilization of technological resources of the industry and of the construction organizations. These methods provide for development of economic-mathematical models and performance of optimization calculations using a computer.

The basic parts of the system are as follows:

(i) Concept for organization of pipeline construction work based on the technology and available technological resources, calculations of the optimum relation between input and surplus volumes, which ensure that the industry runs smoothly;

(ii) Calculation of the productive potential of the industry and of the construction organizations, study of the portfolio of orders, and formulation of the pipeline construction schedules, which are balanced relative to the productive potential;

(iii) Optimization calculations of the organization of construction of each object on the pipeline, determination of the optimum construction time, optimum concentration of resources, structure, disposition and productivity of the production lines, and optimum schedule for putting them into service;

(iv) Development of an optimum schedule for construction of objects along the pipeline;

(v) Development of a system for allocating technological resources under operational management of construction work and forecasting the progress of construction.

The methods developed permit improving the productive potential of the industry and of the construction organizations.

A unified plan for the construction of all main gas and oil pipelines for 1981–1985 is under development.

Soviet builders, fitters and designers scored big successes. They put into operation the gas pipeline Urengoi–Pomary–Uzhgorod 6 months ahead of time.

In September 1983 the Siberian gas, which was transmitted by an underground main line with a length of about 4500 km and pipe diameter 1420 mm, reached the USSR Western border.

The earlier commissioning of this unique construction is of great economic importance. The USSR gas production in 1982 totalled 501 billion cubic meters. When the gas pipeline Urengoi–Pomary–Uzhgorod reaches the design output, the USSR will top the list of gas production countries. But the results may be even better.

The pipeline was built practically in a year, three years earlier than it had been planned. The North "shoulder" with a length of more than 1000 km was built during one winter season, in spite of severe natural and climatic conditions. Simultaneous construction at the Urengoi field of the gas preparation systems, where the capacity of each unit is 20 billion cubic meters per year, gave an opportunity to gradually put into operation the pipeline sections, and, beginning from June, to provide the Urals with additional fuel.

The history of pipeline transport has never seen such a scale and rate of construction.

For example, it took American companies about three years to build the Trans-Alaska crude pipeline with a length of 1280 km and with a pipe diameter of 1200 mm.

The following figures illustrate the scale of work at the gas pipeline Urengoi–Pomary–Uzhgorod. The capacity of ground works was 130 million cubic meters, 2.75 million tonnes of pipes were laid down, and the length of the weld was 2.2 thousand km. The pipeline crossed 150 km of frozen permafrost and 100 km of marsh-ridden areas. The pipeline crossed 560 railways and highways, 800 rivers, lakes and storage lakes.

Subwater crossings were constructed across the following rivers: the Ob', the Kama, the Volga, the Don, the Dnieper. The total length of the subwater pipelines reached 220 km.

The complete cost of the works carried out totalled 15–20 million rubles.

GDR, Poland and Bulgaria met their commitments to the full extent.

2. Experience in Gas Pipeline Construction in the Far North

In the Soviet Union, a great deal of experience has been accumulated in construction of main and industrial pipelines under the conditions existing in the Far North in the permafrost, tundra, and forest–tundra zones.

The Messoyakh, Mostakh, Yuzhno–Solenin, Medvezh'e, and Urengoĭ gas and gas-condensate fields, situated in the permafrost zone, have been assimilated.

Gas is being successfully transported over long distances along the main gas pipelines from these fields. Fields located farther north (Yamburg and the fields on the Yamal Peninsula) will also be assimilated.

The first gas pipelines in the Far North had a diameter 530 and 720 mm (Tas–Tumus–Yakutsk, Solenin–Messoyakh–Noril'sk). However, pipelines recently laid in the permafrost have a diameter of 1420 mm and are pressurized at 7.5 MPa (Medvezh'e–Nadym, Urengoĭ–Nadym, Urengoĭ–Chelyabinsk). The pipelines are being laid underground in embankments and above ground on pile piers.

Analysis of experience in planning, constructing, and operating pipelines under permafrost conditions permits making some generalizations.

It is already clear that the function of the soil foundation of the gas pipeline should not be viewed as a narrow construction problem involving the function of frozen or thawed soil under the action of a foundation with a small load in a definite temperature regime. The problem here is much bigger. To ensure reliable operation of the linear part of northern gas pipelines, it is necessary to create and maintain a new landscape in the entire strip affected by the construction of the pipeline, which is stable over the entire period of pipeline operation. This landscape must be planned based on a comprehensive forecast that takes into account the interrelated changes in the temperature regime of the soil, dynamics of the surface and subsurface flow of water, as well as the physicomechanical properties of the soils.

The correct choice of the technological regime of gas transportation and the method for laying the pipeline plays an especially important, if not leading, role in creating such a landscape.

It is advisable to make maximum use of the underground method of laying gas pipelines at a depth of not less than one meter. This is the cheapest and most reliable method.

Cooling the gas is one of the most effective measures for increasing the reliability of the linear part of pipelines with respect to stabilization of permafrost soils. This method decreases the stresses in the pipe walls and soil erosion to a minimum. In addition, cooling the gas increases the productiveness of gas pipelines.

It is not desirable to cool the gas below 0°C on sections with deep, melted soils. This has a deleterious effect on the operation of the pipeline in underwater crossings due to the possibility of ice formation.

It is desirable to use different temperature regimes for the gas in different geocryological zones.

For zones with continuous permafrost soil, it is recommended that the gas be cooled year-round to a temperature close to the soil temperature (− 1 to 2°C), using refrigeration machines during the summer. In addition, islands of melted soil, encountered along the pipeline route, must be traversed using various methods to counteract swelling and ice formation.

In zones with islands of permafrost soils and farther to the south, the temperature of the transported gas must be no lower than 0°C, based on the conditions required to avoid freezing of the soil around the pipeline. In this zone, it is sufficient to cool the gas using air cooling equipment. In this case, separate islands of frozen soil, when it is not advisable to go around the island, must be traversed by the pipeline either with an allowance for melting or with comprehensive, local measures for protecting the permafrost soils (thermally insulating screens, thermally insulating piles, etc.).

The use of pipes pressurized at 10.0–12.0 MPa is of special interest for gas pipelines in the North. This would permit decreasing the volume of construction work and thereby the zone of destruction, caused by pipeline construction and operation.

Cooling the gas to temperatures close to soil temperatures is a necessary but not sufficient measure to ensure reliable operation of the linear part of the pipeline, since it does not eliminate washing out of the soil used for fill and in the foundation, swelling, flooding, and other undesirable phenomena.

To create a stable landscape in the strip adjacent to the pipeline, it is necessary to undertake an entire complex of structural and engineering-reclamation measures. Such measures primarily include creating drainage and water channels, forming underground and surface flows, building dams that would prevent undesirable percolation along trenches, stabilization of soils against swelling, and pipeline ballasting and anchoring. Physico-chemical methods for technical reclamation of soils improve the structural properties of soils, which is especially important for the northern regions of Western Siberia, where the carrying capacity of sandy loam and loamy soils after melting does not always ensure stable positioning of the pipeline and its sticking in the soil.

Methods of soil fastening by adding cement and oil industry wastes have now been developed and are in use. These methods permit increasing the carrying capacity of soil and improving its anti-erosion properties.

To stabilize the depth to which the soil melts along local pipeline sections and beneath valve foundations and to increase the carrying capacity of pile foundations, thermally insulated piles with forced circulation of a refrigerant have been developed and are being assimilated.

The problem of stabilizing the foundations of northern gas pipelines can be solved more completely by using complex technological methods for improving soil (preconstruction thawing with chemical fastening, antiswelling stabilization with drainage, etc.). An example of such complex measures is the combination of chemical fastening of soils, whose thawing is unavoidable, and a thermally insulated pile for stabilizing the depth of thawing at the base of the fastened soils, currently being used to avoid nonuniform settling of soil at valve blocks.

Some decrease in the thermal effect of the gas pipeline on the foundation soil in order to decrease the depth of thawing or freezing can be achieved by emplacement of thermally insulating screens.

Aside from these engineering–soil-improvement measures, progressive methods for recultivating surfaces affected by construction and for anchoring slopes such as covering with peat-sod mats, sowing grasses, etc., used successfully in water-engineering construction works for anchoring banks of canals, earthen dams, etc. are being introduced into gas pipeline construction and operation. Sections of the tundra that are difficult to recultivate within short periods of time must be replaced by engineered "recultivation" (laying out thermally insulating covers, which provide protection against the destruction of the thermal conditions of the soils where necessary). The development of inexpensive thermally insulating materials based on peats, wood wastes, etc. is of great interest for solving these problems.

Intensive development of gas transportation systems in the northern Tyumen region makes it necessary to develop new technical methods for laying pipe and new operational conditions for northern gas pipelines.

The complexity of construction of northern gas pipelines is illustrated by the Urengoĭ–Chelyabinsk pipeline, which has a diameter of 1420 mm and is pressurized at 7.5 MPa. Out of a total of 1748 km, 1434 km are located within the northern climatic zone and of these, 380 km fall into unassimalated regions.

Permafrost occurs over 470 km and 632 km are flooded. To ballast the pipeline, 199 848 reinforced concrete weights, weighing 4 tons each, as well as 118 704 screw anchors and 1583 hinged pile anchors were required.

The pipeline intersects 5 railroad tracks, 38 automobile highways, 39 rivers more than 30 m wide, 56 rivers less than 30 m wide, and numerous streams and lakes. It was necessary to clear 2180 hectares of forest and 123 hectares of brush. In laying the pipeline, 432 kilometers of temporary logging roads had to be constructed. More than 60% of the line is located in marshy and flooded sections. The complexity index, calculated for this path using the active technique, constituted 3.7.

This shows that the construction of this pipeline is especially difficult. The normal construction time is 36 months.

A unified plan was developed for organizing and managing the construction process. This plan contained the timetable for the linear part, over-all construction plan, transportation scheme, schedules for delivering pipes and other materials, list of equipment and parts for the linear part of the pipeline, instructions for ballasting and anchoring the pipelines, cleaning and testing the pipeline, monitoring of work quality, and organizing communication and machine repair and operation services.

This plan included the engineering charts for the most complex work, as well as charts for operational monitoring of construction–assembly work quality. Surveys of separate sections of the pipeline were also published.

Development of engineering–design documentation specifically related to working conditions facilitates successful pipeline construction in the short times alloted. All construction in the gas industry is performed with project documentation including the plane of the construction organization, formulated by a planning institute and developed by special technological organizations for planning the work.

The work on the Urengoĭ–Chelyabinsk route was performed by large mechanized complex columns (production lines) with completion of preparatory engineering–technological work and adherence to a definite

Table II Duration of Work and the Pace of Production Lines on the Urengoi–Chelyabinsk Pipeline

Sections	Length, km	Total number of production lines, units	Actual number of completed production line-days	Average daily pace along the section per production line, km/day	Actual pace per production line-day
1	2	3	4	5	6
I	283	4	494	1.72/0.43	0.77
II	214	3	404	0.97/0.32	0.53
III	47	1	97	0.96/0.96	0.96
IV	54.8	2	119	0.87/0.44	0.46
V	31.5	1	71	0.44/0.44	0.46
VI	127.7	3	208	1.37/0.46	0.61
VII	93	2	176	0.95/0.47	0.53
VIII	18.3	1	217	0.3 /0.15	0.48
	85.7	1			
IX	30.3	1	360	1.89/0.38	0.49
	145.9	4			
X	132.6	3	198	1.03/0.34	0.67
XI	99	2	368	0.62/0.15	0.58
	126	2			
XII	25	1	295	0.43/0.11	0.56
	140.2	3			
	1800 (1746.8 km for the main line and 53.2 km for the reserve line)	34	302.7		

technological regimen between the production subunits of the line. The duration of the work and the pace of the production lines along the pipeline route are shown in Table II.

It is evident from Table II that the actual duration of the work of a single production column on the pipeline was 89 calendar days, and the actual rate of insulation–laying work performed by a single column, starting from the time the column is positioned on the pipe up to the time the column is disengaged, was 0.6 km day^{-1}.

It should be noted that Siberian construction organizations can greatly increase the length of the construction season and increase the number of working days to 180 by performing the insulation–laying work during the summer along dry sections of the path as well as by organizing preparatory work correctly.

Table III Construction Rates for Linear Part of Separate Pipelines from Western Siberia

Name of pipeline	Diameter, mm	Length, km	Total number of production lines, units	Length of route section constructed by a single production line		Production line output for the winter season (December–April)	
				Average, km	Maximum, km	Average for active production line, km	Average for production line, km
1	2	3	4	5	6	7	8
1. Oil pipeline, Ust'–Balyk, Kurgan Ufa, Al'met'evsk	1220	1844	41	45.0	119	55.7	24.2
2. Gas pipeline, Vyngapur–Chelyabinsk	1420	1431	43	33.3	90	50.4	26.7
3. Oil pipeline, Surgut–Polotsk	1220	1256	33	38.0	92	59.3	33.1
4. Gas pipeline, Urengoĭ–Bhelyabinsk	1420	1748	34	51.4	159	65.6	44.0

The construction rates for the linear part of separate pipelines from Western Siberia are shown in Table III.

Analyzing the indicators of the work performed by the columns, we conclude that due to the increased length of the secitons assigned to a column and, therefore, the smaller number of columns, a 44 km pace was achieved on the Urengoĭ–Chelyabinsk pipeline, while in the first segment of this system (Vyngapur–Chelyabinsk gas pipeline), the pace was 26.7 km.

Analysis of the actual construction schedules and the dynamics of the insulation–laying work during the winter and summer in the North shows that the monthly rate of insulation–laying work during the summer is 6 to 7 times slower than the average winter rate.

One of the main reasons for this situation is the fact that during the summer it is essentially impossible to move along the pipeline route. For transportation and passage of the column, it is necessary to construct hundreds of kilometers of logging roads. The absence of passage along the pipeline route complicates relocation of the construction subunits with the machinery to new sections. The amount of manual labor for insulating the pipelines in locations that are not accessible to passage of mechanical columns increases.

The time required for insulation–laying work was measured for one column in the north during the summer. In April through July, the column completed 1904 man-days of labor and insulated 3.6 kilometers of 1420 mm pipeline with the help of machines and 3.23 kilometers manually. To complete this work, 12.3 kilometers of logging roads were built.

In so doing the following amounts of labor were used:

Mechanized pipeline insulation: 86 man-days;

Manual pipeline insulation: 316 man-days;

Construction of logging roads: 852 man-days;

Relocation of columns to working section: 618 man-days;

Other work: 32 man-days.

From the example presented above, it is evident that only 21% of the labor expended by the insulation–laying column was spent on work the column is intended to perform and, in addition, the rate of mechanized labor to manual labor was about 4 : 1. This situation is typical for summer construction in the North. For this reason, at the present time, efforts are being made to eliminate the seasonal nature (summar season) of main pipeline construction in the North. These efforts are directed toward developing fundamentally new technologies, high-performance multi-terrain machines and transport vehicles, as well as prefabricated, portable roads.

Overall management of the preparatory and construction work for the Urengoĭ–Chelyabinsk pipeline was handled by the Main Regional Production–Management Administration for Western Siberia (GlavterPRU).

To provide for optional management and control of pipeline construction, supervisory sections were created within GlavterPRU for handling production, equipment, and transportation problems. A computer issued daily complete information on the status of the construction.

To control the technological column, Siberian construction organizations created a flexible form of control, which they called the complex high-rate column (CHRC).

In Western Siberia, CHRC was intended to perform a complex of line work on pipelines at a rate of 0.8–1.3 km per working day, 15–20 km per month, and 70–100 km per winter season.

The responsibilities of CHRC included providing for highly efficient operation on all technological stages of linear construction under a unified organizational and economic management. This reduces the pipeline construction time, ensures maximum utilization of machines and machinery, eliminates idle time, and increases the productivity of labor.

The structure of the CHRC is presented in Fig. 1. CHRC performs the entire complex of work on constructing the linear part of the main pipelines: clearing the path, construction of logging roads, hauling out mesh, pipes, bends, anchors and weights, assembly and welding, insulation and placement, trenching and backfilling, loading and unloading operations, ballasting, electrodic protection, blow-through and testing, service and repair for machines and machinery, and housing for laborers, engineers and technicians.

To establish business interrelations and to improve management of construction, a committee of brigade foremen was created in the CHRC.

The committee examined he problems of improving the organization of the labor and the accounting system, quality control problems, problems of timely and high-quality completion of assigned work, etc.

Experience in working with the CHRC system showed a decrease in the organizational–technological tension between interacting subunits.

A unified operational management of the production process permits manuevering resources and taking advantage of resource concentration on a limited operation. Expenditures on servicing the complex decrease, but,

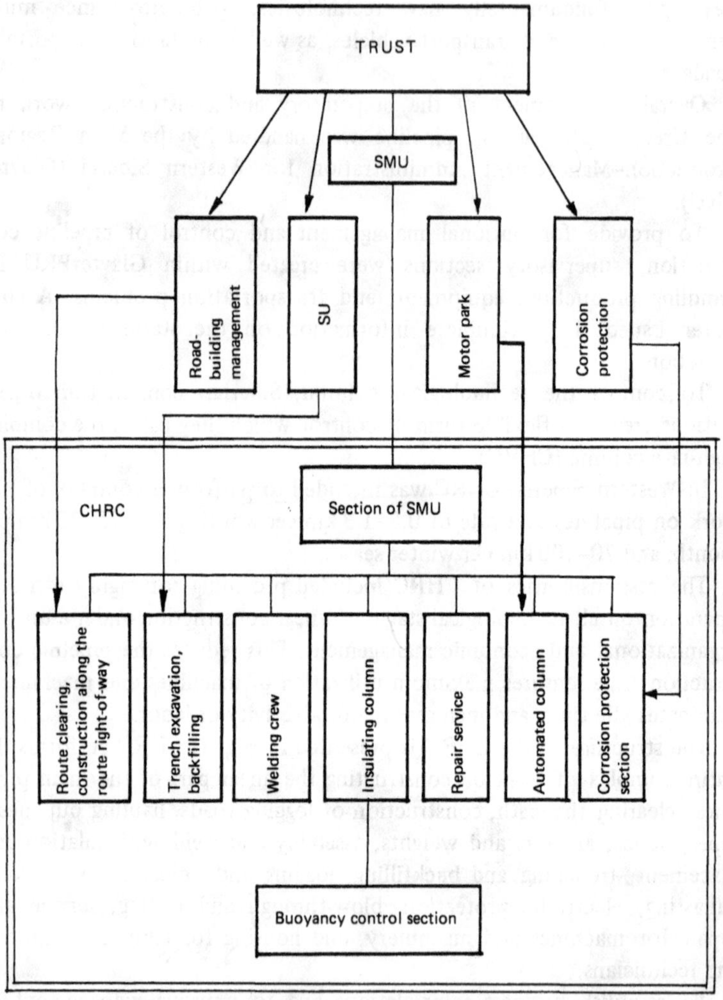

Figure 1 Structure of CHRC.

most importantly, the rate of line work sharply increases and the finished product is delivered: extension and testing of pipeline sections.

Extended observations of the work of four complexes permitted comparing their work efficiency according to two indicators (efficiency of machines and labor)

$$E_m = L/E, \quad (1)$$

$$E_{mL} = L/mL, \quad (2)$$

Table IV Data on Efficiency of Separate Operations, Performed in the Four Complexes Investigated

	Efficiency indicators					
	Ditching		Insulation and laying		Nonrotational welding	
Complex	Em	Et	Em	Et	Em	Et
1	2	3	4	5	6	7
CHRC-1	0.07	0.045	0.153	0.107	0.063	0.029
CHRC-2	0.059	0.035	0.081	0.051	0.034	0.016
CHRC-3	0.062	0.037	0.069	0.043	0.066	0.023
CHRC-4	0.036	0.022	0.073	0.052	0.042	0.013

where L is the length of completed work (km), E is the expenditure of machine time (machine-cm), and mL is the manual labor expended (man-days).

Data on the efficiency of separate types of work performed by the four complexes studied are presented in Table IV.

It follows from the data presented that the work in CHRC-1 was organized most efficienctly.

The characteristics of the complexes analyzed are presented in Table V.

The number of workers per unit differs insignificantly in complexes 1, 2, and 4. CHRC-3 employs twice as many people and 2.2 times more

Table V Characteristics of CHRC

Complex	Number of workers occupied on construction-assembly operations, men	Number of machines used, units	Total machine power, hp	Total accounting cost of the machine stock, thousand rubles
1	2	3	4	5
CHRC-1	258 / 273*	80 / 92*	14 998 / 19 188*	3 740 / 5 204*
CHRC-2	191	87	13 868	3 226
CHRC-3	457	197	37 766	11 162
CHRC-4	225	98	14 324	2 811

*Values including Sever-1 resistance welding crews are given in the denominator.

Table VI Relations Between the Main, Auxiliary, and Service Operations

Type of operation by column	Number of workers	Number of machines and pieces of machinery	Cost of the basic stock
1	2	3	4
CHRC-1			
Main production	1	1	1
Auxiliary production	0.10	0.14	0.01
Service production	1.06	1.93	0.77
CHRC-2			
Main production	1	1	1
Auxiliary production	0.11	0.14	0.03
Service production	0.51	0.80	0.44
CHRC-3			
Main production	1	1	1
Auxiliary production	0.03	0.4	0.01
Service production	0.254	0.400	0.097
CHRC-4			
Main production	1	1	1
Auxiliary production	0.05	0.04	0.01
Service production	0.27	0.76	0.28

machinery, whose power is 2.8 times greater than that of other complexes. Complex 3 actually includes three columns.

It turned out that it was very important to establish an optimum relation between the principle, auxiliary, and service operations, which is characterized by the number of laborers, machines (equipment), and cost of the basic units. These relations are presented in Table VI.

It is evident from Table VI that the best complexes are these in which the numbers of laborers and machines performing auxiliary and service work are comparable to those employed in the principle work.

In recent years, large mechanized complexes (production lines), which perform all the technological operations, leaving behind them kilometers of completed pipeline, have demonstrated that this type of organization is best for pipeline construction in the North. Over the period of a year, large production lines lay 5000–6000 kilometers of large-diameter pipelines.

The basic criterion for choosing the optimal organizational structures was the organizational–technological tension, which describes the ability of a system to preserve its technological unity in the face of the organizational separation of the work process and which is also an indicator of technological specialization.

The greatest gain, according to an integral indicator, was obtained under management by complex trusts, which included welding–insulation and road construction management and specialized subunits for excavation and preparatory engineering work.

The short winter season, which is favorable for linear work, made it necessary to have a closer contact between production resources engaged in construction. Resource concentration, however, reduces the likelihood of achieving a stable rate.

In the middle belt of this country, insulation–laying work is most unstable, but under the conditions in the north, the greatest variations in average daily work rates occur in excavation work, which is highly "sensitive" in Western Siberia.

Preparatory and excavation work with underground pipe-laying under northern conditions dictate the rate of pipeline construction and it becomes necessary to synchronize other operations and to create a unified work organization.

This is possible under unified operation management of the production process, as provided by large-scale mechanized complexes. The advantages of concentrating resources on a limited operation can be best utilized in this case, since it is easier to reallocate resources so as to synchronize separate operations with the critical work.

Increasing the construction capacity of large complexes increases the efficiency of line construction because the cost of servicing the complex decreases.

To increase the organizational–technological reliability of such large subunits, which in the North comprise more than 530 laborers, engineers, and technicians and up to 310 machines and pieces of machinery, a procedure was developed for optimizing the allocation of basic technological and auxiliary machines. The creation of technological and emergency reserve of machines and machinery, following the usual working principle, increased the pipe-laying rate by 10 to 12%.

A column (production line) includes, as a rule, aside from the basic technological subunits of the road construction sections, sections for performing pipeline ballasting operations and for cathodic protection.

This descreases the organization–technological tension between interacting subsections.

Analysis of work performed by pipeline-construction complexes in recent years has shown that, during the winter, under the difficult conditions in the North the output of separate production lines attained

100–150 km of 1420 mm pipe, which corresponds to a rate of about 1 km per day. The welding rate reached 2 km per day, while the insulation rate reached 3–4 km per day.

Since a large mechanized complex includes complex dynamic links between the separate brigades and units of which it consists and has at its disposal a collection of expensive technology, which when standing idle greatly decreases the economic efficiency of the construction process, it is necessary to look for an optimum structure of the complex with respect to different pipeline parameters and specific construction conditions.

To solve such problems, a mathematical model of a large-scale mechanized complex was developed and the required algorithms were constructed.

The construction of main pipelines is characterized by the production-line nature of all types of work performed with extreme specialization of the lower production subunits. The production-line nature of the construction process, as a basic form of organization of pipeline construction work, requires that the large mechanized columns (production lines) be equipped with a large number of high-productivity machines and machinery and they must follow a strict technological sequence with continuous and smooth execution of separate operations. In addition, the construction process must be continuously monitored and efficiently managed.

In constructing the linear part of mainlines, the problem of developing and improving methods for planning and organizing the construction process is especially complicated, since the construction process is highly dynamic and complex and there are many variants for solutions of organizational and technological problems, the work volume and working conditions along the route change continuously, and depending on the season, there is a randomness to the construction process.

Since the network and linear production schedules do not reflect accurately enough the process of constructing main pipelines, the most significant disadvantage is that they represent only one static variant of the model of the construction process, which does not correspond to the diversity of construction conditions and solutions of organizational and technological problems. Some of the factors that are accounted for do not uniquely reflect the conditions occurring in practice. For example, a fixed daily rate is practivally nonexistent in laying main pipelines due to very different working conditions along each section of the route and very diverse weather, geological, topographical, and site conditions.

In their turn, the schedules for delivery of pipes, insulating and other materials, structures, and assembly units to the construction sites do not account adequately for the nonuniformity in the construction process in time and along the route.

For this reason, in recent years, dynamic models, and in particular, simulation models, which reflect the actual construction process most reliably, have been used to develop methods for organizing mainline construction.

Special management has been developed, including management for application of the simulation method.

Development of alternative solutions in multivariant plans of work execution (PWE) greatly decreases the likelihood that the PWE at some stage of construction will no longer be functional due to a sudden change in the construction conditions (for example, a change in season).

Efficient utilization of multivariant PWE in pipeline construction is based on a perfected system for collecting, transmitting, and processing information.

The construction of main pipelines is a complex integral system. Adopting such a system as a starting system (base system) permits keeping it within the limits of pipeline construction as a large, independent, industrial construction industry and simplifies its organizational structure.

The process of constructing main pipelines consists of a complex of simpler process involving construction of separate pipelines, whose objective interrelation and interdependence is determined by the following:

Uniformity of elements and factors (behavior, breakdown, perturbations, ability to react to different actions, etc.);

Unity of the technology, based on the production-line realization of all forms of construction and assembly operations and special work by production lines, column, brigades, and units;

Simultaneous or planned sequential execution of work operations on pipeline routes of different complexity by a definite number of production subunits;

Identical norms for duration of construction;

Since source for material–technological provisions for construction of separate main pipelines.

With such an approach to the structure of the base system, the concept "general system" comprises the planning, technological, management, and organizational systems.

Due to the high rate of construction and the higher technological level of main gas pipelines in the USSR, the gas supply to consumers has been centralized.

Separate gas fields, underground storage facilities, and main gas pipelines and their local subsystems have been combined into a unified gas supply system for the entire country, which has made possible interregional maneuvering of resources and has increased the reliability of the gas supply.

During the 1970's and 1980's the role of gas in the fuel–energy complex of the country increased continuously. The economic value of pipeline gas-transportation systems increased equally. In recent years, they have qualitatively changed and have acquired a new technological and technical appearance.

It is well known that in our country there was an extensive search for methods to increase the efficiency of transporting gas over large distances. It was proposed that pipes with a diameter of 2.5 meters should be built and that gas should be transported in a cooled and liquified form.

Theoretical and experimental studies and technical–economic work have demonstrated that the optimum pipe diameter is 1420 mm and the optimum working pressure for transporting gas along such pipelines is 7.5 MPa and possibly as high as 10.0–12.0 MPa.

The optimization studies included the complex of technological and construction factors. A fundamentally new contribution was the inclusion of the actual working capacity of the pipes and the ability of the pipe industry to provide the required level of reliability.

The conversion to gas pipelines with a diameter of 1420 mm, first undertaken in our country, makred a technical revolution.

The productiveness of gas pipelines with a diameter of 1420 mm pressurized at 7.5 MPa is 3–3.5 times higher than than the producitiveness

Table VII Labor Expended per 1 km of Pipeline, Man-Days*

Region (locality)	Standard pipeline diameter, mm 1000	1200	1400
Plain–hilly	476	578	766
Swampy in European part of the country	605	736	977
Deserts and semideserts	510	618	818
Mountain	667	809	1071
Swampy in northern regions of the country	722	879	1164
Northern	812	990	1310

*Labor and wages per 1 km of pipeline increase by 5 to 8% for construction of cable or radio relay communication lines.

Table VIII Wages per 1 km of Pipeline (in Rubles)

Region (locality)	Standard pipeline diameter, mm		
	1000	1200	1400
Plain–hilly	5 035	6 115	8 094
Swampy in European part of the country	7 047	8 572	11 370
Deserts and semideserts	6 465	7 844	10 384
Mountain	8 492	10 286	13 617
Swampy in northern regions of the country	10 738	13 027	17 280
Northern	14 657	17 823	23 572

of previously constructed pipelines with a diameter of 1220 mm pressurized at 5.5 MPa, the savings in metal per 1000 cubic meters of gas constituted 20%, the unit capital investment decreased by 30%, and the productivity of labor in the gas increased by a factor of 3.

One of the most important indicators of the efficiency of constructing 1420 mm pipelines is the decrease in labor required to transport 1000 cubic meters of gas. Tables VII and VIII present the cost of labor and wages for the work required to lay 1 kilometer of pipe with diameters of 1020, 1220 and 1420 mm. The indicators presented are differentiated according to six construction regions, which differ by climate topography, variation of the regional cost-determining factors, and the volume of and technology used in the construction and assembly work.

The conversion to labor costs and wages for 1 million rubles spent on construction–assembly work (Tables IX and X) is done via the capital investment in new construction per 1 kilometer of pipeline.

Table XI presents the average annual output per laborer, performing construction–assembly and ancillary work, with an average of 265 working days per year.

Table IX Labor Costs per One Million Rubles of Estimated Cost of Construction–Assembly work, Man-Days

Region (locality)	Standard pipeline diameter, mm		
	1000	1200	1400
Plain–hilly	5051	4533	4227
Swampy in European part of the country	4794	4307	4022
Deserts and semideserts	5691	5107	4748
Mountain	5856	5249	4887
Swampy in northern regions of the country	4793	4308	4014
Northern	4390	3947	3770

Table X Wages per Million Rubles of Estimated Cost of Construction–assembly Work, in Rubles

Region (locality)	Standard pipeline diameter, mm		
	1000	1200	1400
Plain–hilly	52 374	47 007	43 833
Swampy in European part of the country	54 698	49 194	45 856
Deserts and semideserts	70 763	63 312	59 125
Mountain	73 044	65 363	60 680
Swampy in northern regions of the country	69 807	62 650	60 158
Northern	77 378	69 598	66 445

The indicators included expenditures on the following combined groups of work operations:

Preparatory and earth-moving;

Welding–assembly (welding of rotating and nonrotating joints, assembly of collars, valve blocks, floating-bridge crossings;

Insulation–placement, ballast;

Blow-through and testing;

Laying pipeline across water barriers and roads;

Installing cathodic protection;

Other (constructing temporary structures and temporary and permanent roads, expenditures on repair and technical servicing of machines, relocation of the construction–assembly organizations, preparation of construction, as well as expenditures related to winter work).

Labor costs, referred to the productiveness of gas pipelines, show that

Table XI Average Yearly Output Per Worker, Performing Construction-assembly and Ancillary Work (for Average Yearly Total Working Time of 265 Days), in Rubles

Region (locality)	Standard pipeline diameter, mm		
	1000	1200	1400
Plain–hilly	52 342	58 319	62 544
Deserts and semideserts	46 449	51 770	55 679
Mountain	45 146	50 368	54 101
Swampy in northern regions of the country	55 159	61 370	65 852
Northern	60 216	66 971	70 119

in creating a transportation capacity of 1000 cubic meters using 1420 mm pipe, 1.5 times less labor is used than for 1220 mm pipe.

Wages, referred to the productiveness of gas pipelines, for developing a transportation capacity of 1000 cubic meters using 1420 mm pipe are 1.5–1.6 times lower than for 1220 mm pipe.

The indicators in Tables IX and X indicate the decrease in labor costs and wages per 1 million rubles of the estimated cost of construction-assembly work accompanying the increase in pipe diameter.

At the same time, the average yearly output per worker (Table XI) in construction of large diameter pipelines is much higher.

In constructing pipelines with a diameter of 1420 mm, large savings are achieved due to the introduction of new technology and decrease in construction time. This can be demonstrated for the construction of the Urengoĭ–Chelyabinsk pipeline.

Completion of sections of the Oreng–Chelyabinsk pipeline ahead of schedule produced an additional 5761 million cubic meters of gas in 1979.

The total economic impact (E_{tot}) consists of the following:

1. Economic impact of unit costs due to introduction of new technology (E);

2. Economic impact from decrease in construction time (Et), which consists of:

Impact of decrease in constant standard expenses of construction organizations due to the decrease in construction time (Es);

Impact due to pipeline operation during the additional period of time resulting from the completion of the pipeline ahead of schedule (Ef).

$$Et = Es + Ef. \tag{3}$$

Then

$$E_{tot} = E + Es + Ef \tag{4}$$

or

$$E_{tot} = E + Et. \tag{5}$$

The impact of the decrease in the constant standard expenses of construction organizations is determined from the equation:

$$Es = H\left(1 - \frac{T_2}{T_1}\right). \tag{6}$$

The impact due to pipeline operation during the additional period of time resulting from completion of pipeline ahead of schedule is determined from the equation:

$$Ef = Eh \cdot F(T_1 - T_2). \tag{7}$$

This equation is used because the organization using the pipeline does not keep a separate account of the financial result by succession of the

Table XII Data for Calculating Economic Efficiency Due to Introduction of New Technology and Decrease in Construction Time of the Urengoĭ–Chelyabinsk Gas Pipeline

Indicators	Notation	Units	Magnitude of indicator
Duration of pipeline construction according to plan	T_1	Year	2.0
Actual duration of pipeline construction	T_2	Year	1.417
Standard fixed expenses for variant with construction duration T_1	H	1000 rubles	57 296
Cost of basic stock, put into operation ahead of schedule	F	1000 rubles	682 884
Standard coefficient of investment efficiency for gas industry	Ef		0.12
Economic impact according to reduced costs from introduction of new techniques and progressive technology (CHRC, AR-401, SEVER setup, pipe cementing)	E	1000 rubles	14 772

gas-pipeline system. Data for the calculations are presented in Table XII.

Substituting the data in Table XII into the equations, we obtain:

$$Es = 57\,296\left(1 - \tfrac{1.417}{2.0}\right) = 16\,730 \text{ thousand rubles},$$

$$Ef = 0.12 \times 682\,884(2.0 - 1.417) = 47\,774 \text{ thousand rubles},$$

$$Et = 16\,730 + 47\,774 = 64\,504 \text{ thousand rubles}.$$

The savings due to the decrease in construction time for the Urengoĭ–Chelyabinsk gas pipeline amount to 64.5 million rubles.

Substituting the data in Table XII and from the calculation into the equation, we obtain:

$$E_{tot} = 14\,772 + 64\,504 = 79.3 \text{ million rubles}.$$

Thus the savings resulting from the introduction of new technology and the decrease in construction time for the Urengoĭ–Chelyabinsk gas pipeline amount to 79.3 million rubles.

With respect to the normal construction time, the savings due to completion of the pipeline ahead of schedule and introduction of new technology amounted to 150 million rubles.

All standards for gas pipelines in the new 1420 mm class pressurized at 7.5 MPa have been developed. The technology and construction organization required for this class, as well as the means for complex mechanization, which permitted increasing the level of mechanization to 99.7%, i.e., completely mechanizing the construction of pipelines in this class with a high degree of automation of welding work, have also been created.

Main gas pipelines are extremely expensive structures requiring large quantities of metal. Recall that a 1000-kilometer section of pipeline made of 1420 mm pipes costs up to 1 billion rubles and requires 650–700 thousand tons of steel. Under these conditions, optimal planning of the pipeline course and optimal solutions of construction problem permit enormous savings in facilities, metal, and labor costs.

It is enough to say that decreasing the length of a 1420 mm pipeline by only 1% can reduce the capital investment by up to 10 million rubles, consumption of metal by up to 6500–7000 tons, and labor costs by up to 10 000 mandays.

3. Mainline Construction Operations

3.1. Earth-Moving Work

The volume of earth moved in oil and gas construction sites constitutes 500–600 million cubic meters per year.

In 1980, 546 million cubic meters of earth were moved. Of this volume, 349 million cubic meters were moved during pipeline construction: 156 million cubic meters for digging trenches and 193 million cubic meters for route layout and trench filling.

Surface construction involved 101 million cubic meters and 96 million cubic meters were moved in the course of road construction.

Earth-moving operations in constructing underwater crossings of main pipelines along the linear part likewise involve about 20 million cubic meters. Here, 3 million cubic meters are excavated at depths up to 10 m, 12 million cubic meters at depth from 10 to 15m, and 5 million cubic meters from 15 to 25 m and more.

The ground is recultivated in a strip 12 m wide and, in this case, the average thickness of the layer scraped off is 0.3 m. In constructing 10 000 kilometers of main pipeline per year (approximately 40% of which are subject to recultivation), 24.8 million cubic meters of earth will be moved. The area of the fertile layer per 1 km of 1420 mm pipeline is, on the average, 1.26 hectares.

To increase the rate of earth-moving operations and the efficiency with which the earth-moving machines are utilized, efficient differentiated treching methods, using different complexes of digging machines depending on the soil conditions, are used in main pipeline construction.

The first method involves digging trenches with single-bucket excavators and bulldozers–rippers. The highest efficiency in using this method is achieved in digging heavy, frozen, and rocky soils with preliminary ripping, as well as in digging trenches in frozen swamps.

The second method makes use of rotary trench excavators and bulldozers for digging trenches. This method is recommended for continuous excavation of trenches using rotary excavators at a high rate under large-diameter pipelines in class I–III soils.

In the third method, rotary excavators are used to dig the trench. The rate of trench digging using this technology increases practically proportionally to the number of excavators in the complex (the rate reaches 3–4 km/day).

Trenches for main pipelines are excavated using different digging machines. This is determined by the change in soils, especially in Western Siberia, where swamps that do not freeze even in the coldest weather alternate with sandy crests which freeze to the total depth of the trench and with sections containing permafrost. One of the most important areas of mechanization of earth-moving operations in pipeline construction is the development and application of high-productivity rotary trench excavators.

Rotary excavators manufactured in the Soviet Union by SKB "Gazstroĭmashina" are distinguished by their high technical performance. They are capable of digging frozen soils both in the middle belt of the country and in the northern regions. This is achieved primarily by the staggered arrangement of the teeth on the buckets of the rotor, which permitted decreasing their overall number by a factor of 3–3.5 as well as the energy expended on the cutting process. Use of a wear-resistant cutting tool with an efficient geometry and teeth reinforced with hard-alloy blades, whose durability exceeds the durability of teeth with hard-facing by factors of 10, also facilitate successful excavation of frozen soils. Rotary excavators have power plants with high absolute and specific power capacity, providing high linear productivity.

They can move over large distances without roads and in rugged terrain due to a reliable track and good weight distribution. The excavators have a reliable and strong construction and quite high ergonomic indicators. The machines have a semitrailer arrangement, in which the working organ is supported by the wheel base both in transport and working positions and not on the skid base. In addition, the mass of the prime mover and of the entire excavator is smaller, the base tractors can be used with minimum conversion, the cabin can be placed closer to the working organ, and the reliability and lifetime of the track are higher.

For the first time in world practice, a diesel-electrical drive is used for rotary trench excavators. The technical characteristics of a series of rotary trench excavators are presented in Table XIII.

Excavation of permafrost and seasonally frozen soils is especially complicated. SKB "Gazstroĭmashina" developed and manufactures serially at the MÉMZ Minneftegazstroĭ plant the rotary ÉTR-254 trench excavator and its modification ÉTR-254-02, shown in Fig. 2. These excavators have a greater digging depth for developing trenches beneath main pipelines with diameters from 720 to 1620 m in soils up to class IV inclusively and in soils frozen to the depth of the trench. The metal parts in the excavators consist of steels that ensure reliable operation at low temperatures. The excavator operates at ambient air temperatures from $-40°C$ to $+40°C$. The base of the excavator is a special base using units from the series K-701 tractor and the engine power is 300 hp. The dimensions in the transport position are: length 13 450 mm, width 4230 mm, and height 4770 mm. The excavator weighs 41 tons. The excavator can cut a trength to a depth of 2.5 and 3.0 m and width at the bottom from 1.8 to 2.4 m. With the use of backslopers, the maximum width of the trench at the top is 3.8 m. The technical productivity in class I and II soils is 1200 cubic meters per hour and the working speed ranges from 20 to 509 meters per hour (32 speeds). The average specific pressure on the soil is 0.73 kg cm^{-2}.

The excavator is equipped with a comfortable cabin and has a pre-operational diesel heater. All this creates good conditions for operating machines under northern conditions. The use of ÉTR-254 excavators on pipeline courses, including laying 1420 mm pipe, had demonstrated its high level of performance.

ÉF-131 and ÉF-251 cutting excavators with a novel construction have been developed to cut slit trenches in frozen and permafrost soils. Narrow slit trenches, cut by the excavators, can be used for laying cables and small-diameter pipelines, as well as for digging trenches under main pipelines using explosions or earth-moving machines. Figure 3 shows an ÉF-251 cutting excavator in the transport position and Fig. 4 shows an ÉF-134 excavator in operation.

Table XIII Technical Characteristics for Rotary Trench Excavators

Parameters	ÉTR-204	ÉTR-223	ÉTR-224	ÉTR-231	ÉTR-253A	ÉTR-254
1	2	3	4	5	6	7
Profile of excavated trench	Rectangular with banks with slope 1 : 0.3 at depth 1.3 m	Rectangular with banks with slope 1 : 0.32 at depth 1.4 m	Rectangular with banks with slope 1 : 0.32 at depth 1.4 m	Rectangular with banks with slope 1 : 0.32 at depth 1.5 m	Rectangular with banks with slope at depth 1.2 m	Rectangular with variable banks with slope from 1 : 0.27 to 1 : 0.58 at depth 2 m
Dimensions of trench, m:						
a) maximum depth	2.0	2.2	2.2	2.3	2.5	2.5
b) width without banks	1.2	1.5	0.85	1.8	2.1	1.8, 2.1; 2.4
Power:						
a) engines, hp	160	160	160	250	300	300
b) generators, kW	–	–	–	200	200	–
c) electric motor for rotor, kW	–	–	–	100	125	–
d) electric motor for conveyor, kW	–	–	–	17 (two)	17 (three)	–
Excavation efficiency in category 1 soils, $m^3 h^{-1}$	650	650	600	800	1200	1200*
Velocity during excavation, $m h^{-1}$	10–300 (Continuous)	10–300 (Continuous)	10–300 (Continuous)	38–224 (8 speeds)	20–350 (Continuous)	20–509 (32 speeds with shifting in motion)

Transport velocity, $km\ h^{-1}$	1.58–5.22	1.58–5.22	1.58–5.22	1.34–3.68	2.3 and 6.0	0.48–5.6
Rotor diameter relative to tips of teeth), mm	3550	3830	3830	4150	4500	4350
Rate of rotation of rotor, rpm	9.6 and 7.8	9.0 and 7.2	9.0 and 7.2	7.9	7.4	7.66
Strip width, mm	800	800	800	1000	1200	1200
Velocity of strip, $m\ s^{-1}$	5 and 4	5 and 4	5 and 4	5	4.9	5 and 3.5
Average pressure on soil, $kg\ cm^{-2}$	0.6	0.7	0.7	0.65	0.9	0.67
Load on rear wheels in transport position, tons	–	–	–	10	19	11
Dimensions in transport position, m:						
a) length	11.1	11.5	11.1	12.8	13.4	13.4
b) width (without conveyor)	3.2	3.2	3.2	3.2	3.7	3.5
c) height	4.2	4.2	4.2	4.4	5.0	4.8
Weight, tons	30.0	32.8	31.5	43.0	59.8	41.0

*Productivity in category 1 and 2 soils.

Figure 2 ÉTR-254-02 rotary trench excavator.

Figure 3 ÉF-251 cutting excavator in transport position.

Figure 4   ÉF-131 cutting excavator in operation.

The prime movers of the cutting and ÉTR-254 excavators are highly standardized. The base of the ÉF-251 excavator is a special base consisting of units from the K-701 series tractor. The engine power is 300 hp and the dimensions are: length 11 650 mm, width 3200 mm, maximum height 5500 mm, cutter diameter along teeth 3800 mm, and excavator mass 29 tons. Cutting excavators were developed by the SKB "Gazstroĭmashina," and are manufactured by the MÉMZ plant of Minneftegazstroĭ. The excavator cuts a trench with a depth up to 2.5 m and width 0.27 m. The technical performance in frozen soils is 70 cubic meters per hour and the speed in operation varies from 20 to 509 $m\ h^{-1}$ (32 speeds).

Single-bucket excavators, including excavators for northern operations based on rubber–metal track with bucket capacity of 1.25 cubic meters and other models, have been developed for use in swampy sections of the pipeline course, where rotary excavators cannot be used.

The exacting requirements for protecting the environment, making sure that the insulation of the pipeline is not damaged during placement of the pipe and backfilling of the trench, made it necessary to create a series of new machines. Special high-productivity rotary machines were developed for recultivating fertile soils, since the area of the regenerated arable lands attains 600 000 hectares per 1000 kilometers of large-diameter pipeline course.

A TR-351 rotary trench filler with a novel deisng, shown in Fig. 5, was developed. The trench filler is intended for continuous filling of the trench bottom with pulverized soil and covering the pipeline placed in the trench. In so doing, it is ensured that there is no damage to the pipeline insulation, a smaller right-of-way is required, the machine does not track on previously laid strings when a parallel pipeline is being constructed, and the work of the operator is made easier compared to work with a bulldozer. The machine works thawed and frozen soils at ambient air temperatures from $-40°$ to $+40°C$. The prime movers of the trench backfiller and rotary ÉTR-254 excavators are standardized. The dimensions are: length 8950 mm, width without carrier 4960 mm, height 3400 mm, rotor diameter 2420 mm, excavator mass 36.5 tons, and average specific pressure on the soil 0.8 $kg\ cm^{-2}$.

The trench backfiller has a special base utilizing the K-701 tractor with a 300 hp engine. The working speed varies from 35 to 366 m/h (16 speeds). The technical performance in category I soils is 1200 $m^3\ h^{-1}$. Tests and operation on pipeline courses have demonstrated the higher reliability of the trench backfiller.

Figure 5 TR-351 rotary trench backfiller.

3.2. Pipeline Welding

A large amount of welding is done on main pipeline courses. About 2.5 million joints are welded annually and the total length of welding seams is 9000 km. One-half of all joints are welded by automatic welding.

Failure of welded joints in most (70%) cases involves manual welding. In the case of automatic welding, joint failure also primarily involves internal manual welding. In the Soviet Union, pipeline welding is performed in two stages: welding of pipes in two- or three-pipe sections with a length of up to 36 meters on pipe-welding platforms and welding of sections into a continuous string along the course.

Until recently, nonrotating joints were welded primarily by the continuous-line method of separated and group manual welding.

Rotatable joints for large-diameter pipes are welded on specialized automatic welding platforms. The only nonmechanized operation that remains here is welding the root of the seam inside the pipe. At the present time, automatic two-sided submerged arc welding on the BTS-71 platforms are used for welding 720–1020 mm pipes; BTS-142 platforms are used for welding 720–1420 mm pipes in two-pipe sections; and, BTS-143 and BTS-142V platforms are used for welding three-pipe sections with diameters up to 1420 mm. Two-sided automatic submerged arc welding technology with complete mechanization of preparatory and assembly operations as well as monitoring of the quality of the welded joints have been realized on these platforms.

A characteristic feature of the technology of two-sided automatic submerged arc welding is that by changing the grooves of pipes (increasing the root face and decreasing the groove angle) arriving from the factory, it is possible to eliminate manual welding completely and to decrease the amount of metal used per joint by a factor of 3 to 4.

Elimination of manual welding practically eliminated internal defects in a joint and greatly increased the quality of the welded joints.

The BTS-143 welding platform, illustrated in Fig. 6, includes four stands for working the rims, welding three-pipe sections, and inspecting completed sections. The maximum power used is 350 kW. Mechanized assembly of pipe into sections up to 36 m long, two-sided automatic submerged arc welding of 1420 mm pipes with walls up to 20.5 mm at ambient air temperatures varying from $-40°C$ to $+40°C$, as well as inspection of the finished weld joint are performed on this platform.

The pipe rims are worked by special stands at two working locations, situated in enclosures. The end pipes in the three-pipe section are prepared for automatic welding from one side and for internal welding from two sides.

Figure 6 BTS-143 welding platform.

The first pipe entering the stand for welding three-pipe sections is moved in such a way that it moves onto a bar carrying the aligner and the welding head and the pipe end being worked ends up between the clamps of the aligner. The second pipe is fed onto a roller conveyor until it touches the first pipe. The pipes are aligned and the first external layer of the seam is welded with the outer head. After this, the aligner clamps are released and the bar is moved inside the section in such a way that the tip of the electrode wire of the interior welding head coincides with the plane of the joint. The second external layer of the seam is welded simultaneously with the internal layer.

The operator follows and controls the position of the electrode during internal welding remotely, for which purpose there is a special electromechanical tracking system.

The pipes are rotated with the help of a power-driven roller operator.

The welded two-pipe section is transferred to the stand for welding three-pipe sections. By this time, a single (third) pipe is already located on the bar of the three-pipe section. The equipment and technology for welding the third pipe are similar to that used for welding the two-pipe section.

The finished three-pipe section is transferred by roller conveyors along the production line to the inspection and repair stations. Detachment of separate pipes and sections, stacking, and removal from the work stations are performed hydraulically.

The inspection stand has two work stations, which makes it possible to inspect the quality of both joints simultaneously. Each work station has it own mobile laboratory. The X-ray equipment is mounted on a trolley, which makes it easier to move the equipment inside the pipe and to place it on the joint. The operator controls the irradiation process from within the enclosure. After inspection, the section is stacked.

On the BTS-143 platform, three three-pipe sections are welded per hour: 6 joints for 1420 mm pipe with walls 17 mm thick, i.e., it is possible to weld up to 60 high-quality joints per shift on this platform.

In contrast to its analogs abroad, welding on the BTS-143 pipe welding platform does not require the presence of a welder within the pipe during internal welding, since the process is done remotely. All work stations have individual enclosures, which permits working in any weather during the entire year. The equipment on the platform, which weighs 195 tons, can be placed into a standard portable enclosure with a 12-meter span.

For a long time it was not possible to solve the problem of automatic welding of "nonrotatable" joints of pipe sections while stringing pipe along the pipeline route. In recent years, automatic powdered-wire, gas-arc, and resistance welding units have appeared on mainlines. The E. O.

Figure 7 Sever-1 contact welding complex.

Paton Institute of Electric Welding in partnership with the organizations of Minneftegazstroĭ developed the Sever-I complex of machines for resistance welding of 1420 mm pipelines. This method insures high-productivity and high-quality welding.

High-productivity is a result of the high (relative to arc methods) power input to the joint simultaneously over the entire perimeter of the pipe at a single working position, combining the assembly and welding operations, total automation of the welding process, as well as elimination of the need for moving the heat source or rotating pipes during welding.

Flash resistance welding requires unit power input of 0.5–0.7 kW cm^{-2} and consumes 0.02–0.03 kW · hr cm^{-2} of electricity. The welding time for a single joint was reduced to 90–120 seconds independent of its diameter.

High-quality pipe joints are achieved with resistance welding due to the total automation of the process, which is controlled by a special system and corrects accidental changes in the technological welding parameters and introduces adjustments for random changes in the geometric dimensions of pipes and characteristics of welding machines.

Resistance welding guarantees that absolutely gas-tight seams are obtained. Defects such as hardening cracks, bubbles, and shrink holes, which are the most common reasons for the failure of welded seams, do

Figure 8 In-pipe unit.

not occur in resistance welding. In addition, ther internal stress level in transverse girth seams are several times lower with resistance welding than with electric arc welding.

The high quality of joints welded by the resistance method is confirmed in practice. Resistance welding has been used along more than 26 000 kilometers of main pipeline without even a single case of joint failure.

The Sever-I resistance welding complex, shown in Fig. 7, is based on a fundamentally new technical foundation. Its main assembly unit, the resistance welding machine, is situated inside the pipeline and moves along it from joint to joint in a self-propelled manner, guided by the pipeline. This not only increases maneuverability, but also decreases time lost on relocation when welding nonrotatable joints. This creates special conveniences when working under conditions without roads. The intrapipe unit aligns the pipes (Fig. 8), welds, and removes internal burrs. It has a diameter of 1350 mm and a length of 11.5 mm, and weighs 38 tons.

Welding requires a 1000 kW electrical power plant. The electrical power plant moves along the pipeline and after adding on the next pipe or section it is connected to the intrapipe machine through a bar by a special detachable plug. The unit removing the external burrs moves along the pipeline or is transferred by the pipe-layer from joint to joint.

The complex includes a unit for cleaning the ends of the pipes prior to welding as well as pipe-layers for feeding and assembling the next pipes and sections.

The nominal welding current is 100 000 A and the nominal secondary voltage is 6 V. The drive mechanisms for clamping, flashing off, and setting is hydraulic. The setting force is 400 tons.

The large amount of heat introduced into the welding joints reduces the welding time for a single joint to 4 minutes. The complex can weld up to 8 joints per hour. Thus 50–65 joints can be welded per shift, which with the use of three-pipe sections permits laying more than 2 km of pipeline with a diameter of 1420 mm. The output per worker is 3–4 times higher than that achieved with manual arc welding.

The complex is operated by 12 men, and it frees 36 qualified welders.

Resistance welding by the Sever-I complex on northern courses has shown that automated control of welding and inspection ensures stable production of high-quality joints.

The successful experience with resistance welding of gas pipelines with a diameter of 1420 mm under northern conditions, performed for the first time in the world, opens up new possibilities in the construction of main pipelines both with respect to the rate of construction and achievement of high quality and reliability.

There is every reason for resistance welding to become the dominant welding technology in pipeline construction. It is projected that in 1981–1985 the volume of resistance welding will reach 50% of all joints on small-diameter pipelines and up to 30% on large-diameter mainlines. To achieve such results, over the years indicated more than 100 resistance welding machines will be manufactured. In recent years, resistance welding of pipes with a wall thickness of 26 mm has been successfully tested, which makes this method promising for application to high-pressure pipelines.

Foreign companies have shown great interest in using the method of resistance welding pipes, developed in the USSR, for welding marine pipelines. Licenses have been sold in the USA and in Japan.

Approximate calculations show that the use of resistance welding in marine construction and for laying underwater marine pipelines will permit using pipe-laying barges whose displacement is approximately 4–5 times smaller than that of barges used at the present time or quadrupling the productivity of existing pipe-laying barges.

The continuous-line method of separated and group manual welding is widely used in the construction of main pipelines. With this type of work organization, several joints are welded at the same time and a unit of 2–4 welders welds each layer or each section of the seam.

The E. O. Paton Institute of Electric Welding, VNIIST and the Kiev affiliate of SKB Gazstroĭmashina developed a complex for gas-arc welding of nonrotatable joints of 1220 and 1420 mm pipes called Duga and a machine for automatic welding of nonrotatable joints with powdered-wire welding. In powdered-wire welding, the girth seam is welded with forced formation of the seam metal by a copper slider, which permits welding under more intense conditions (welding current up to 500 A) and which therefore provides high productivity. A special self-shielding powdered wire provide high quality welding joints without additional shielding of the weld pool, and this is a big advantage. The new technology has been successfully tested on a pipeline route and increases productivity by a factor of 2–3 compared to manual arc welding and provides high-quality joints.

3.3. Insulation and Pipe Laying

In the Soviet Union, the technology in which the pipe-laying is combined with cleaning and insulation is most widely used. Due to the high rigidity of a 1420 mm pipeline, when the pice is placed on the pipe-layer hooks, about 200 meters of pipe weighing 180 tons, including the weight of the insulation and cleaning machine, are suspended. The pipe-layers are the main weight-lifting machines used in mainline construction. Progressive solutions to technical problems that have been incorporated in the construction of new pipe-layers include the following: possibility for combining work operations, lowering the hook and crane arm with the driver only, automatic disconnection of the hoist brakes, small number of control levers requiring small forces for engaging them, a system for monitoring the load stability and load, and more powerful and more reliable caterpillar tracks. These features ensure that these machines have better performance than previously manufactured series pipe-layers.

In recent years, a series of pipe-layers has been developed and is manufactured, including the heavy TG-502 pipe-layer (Fig. 9). This pipe-layer, intended for laying pipe with diameters 1220–1420 mm into a trench, for accompanying the cleaning and insulating machines and performing different lifting–transporting operations on the welding platforms, in assembly or valve units, etc., was developed by SKB Gazstroĭmashina based on modification of the commercial T-330 tractor for pipe-laying work. The Strelitamak plant, which manufactures construction machines for Minstroĭdormash, manufactures the TG-502 pipe-layers.

The nominal weight-lifting capacity of the pipe-layer is 50 tons and the stability moment is 110 ton-meters. The maximum load on the carrying hook when operating in the insulating–laying column is 70 tons. The

Figure 9 TG-502 pipelayer.

lifting height of the hook is 6.3 m and the lowering depth is 2.0 m from ground level when the hook is projected from the tipping axis by 1.5 m. The engine power is 270 hp, and the weight is 59 tons. The average pressure from the left track on the soil with a load of 50 tons on the hook and using the entire stability moment does not exceed 3 kg cm^{-2}.

Compared to the CAT-504 pipe-layer, the Soviet model has a number of advantages:

smaller number of levers for controlling mounted equipment and smaller forces required to move them;

safer operation of the load-lifting mechanism;

better transportability;

possibility for combined operation of load-lifting and crane-lifting hoists;

drive for mounted equipment has been converted to hydraulic operation.

A heavy pipe-layer with a novel design (TG-634) has been developed. This pipe layer has a nominal weight lifting capacity of 63 tons and a stability moment of 170 ton–meters. The maximum load on the load carrying hook when operating in an insulation–laying column is 88 tons. The lifting height of the hook is not less than 6.3 m and the lowering depth of the hook from ground level is about 2 m when projected from the tipping axis by 1.5 m. The engine power is 160 hp and the weight is 650 tons. The average pressure of the left track on the soil with a load of 63 tons on the hook and using the entire stability moment is 344 Pa (3$1 $kg\,cm^{-2}$).

The TG-634 pipe-layer is shown in Fig. 10. Its distinguishing feature is the variable caterpillar track, which with good transportability decreases the unit consumption of material from 0.46 tons/ton–meter to 0.38 tons/ton–meter and saves 13 tons of steel per machine, since the stability moment is decreased not only by increasing the mass of the pipe layer, but primarily by increasing the distance from the center of gravity of the machine to the left tipping edge. In the transport position, the machine has the same dimensions, while i operation, the machine extends by 1000 mm.

The TG-634 pipe-layer has a number of advantages over the CAT-594 pipe-layer: a 50% higher stability moment, better transportability using railroad transportation, smaller number of levers for controlling mounted equipment, possibility for combined operation of the crane arm and the hook, and the crane arm and the hook are lowered only by the power plant.

To lay concreted mainline pipe with diameters up to 1020 mm on river floodplains and in swamps, as well as for repaining pipeline, it was necessary to develop a bridge crane-type pipe-laying machine. The weight-lifting capacity of the machine is 50 tons. The maximum and minimum distances

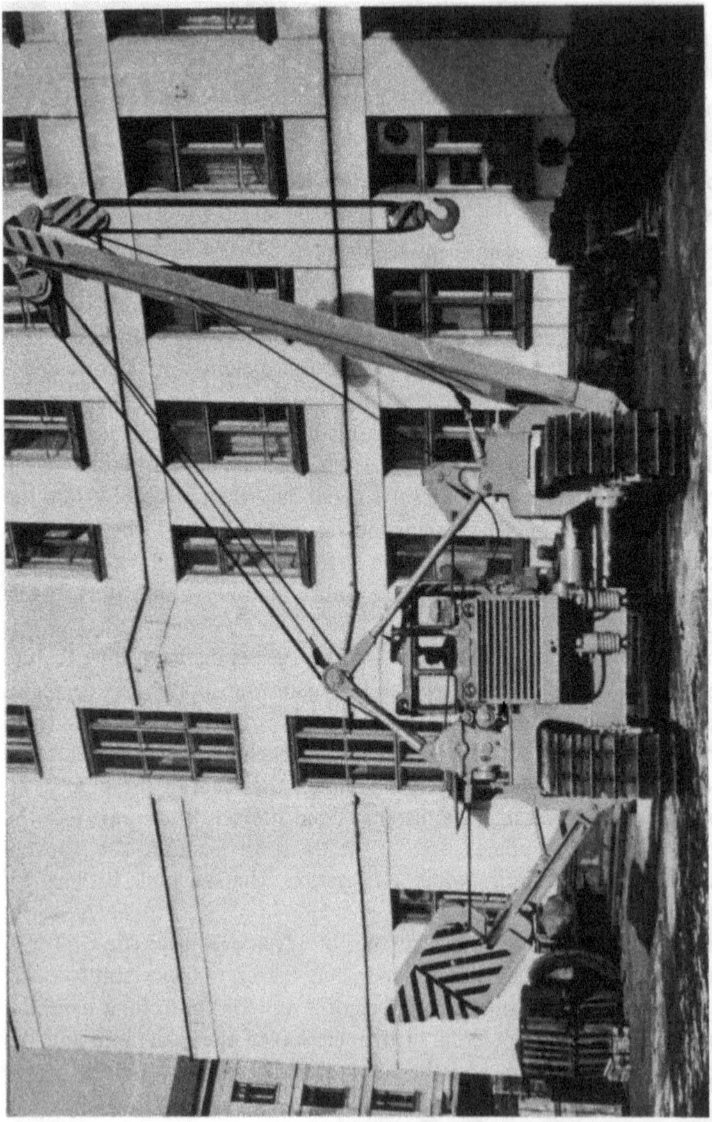

Figure 10 TG-634 pipelayer.

Table XIV Short List of Characteristics of Pipe Layers Used in Pipeline Construction

No.	Pipe layer	Weight-lifting capacity and load moment, tons × ton-meters	Base tractor
1.	TG-61	6.3 × 16	DT-75P-C3
2.	TG-62 (for swamps)	6.3 × 16	DT-75P-C3
3.	TG-124	12.5 × 34	T-130
4.	TG-201	20 × 50	T-130
5.	TE 560M	35 × 75	D-804M
6.	TG-502	50 × 125	TT-330
7.	MM-631* (expanding tracks)	63 × 240	TT-330
8.	TG-801*	80 × 160	TT-500
9.	MTM-501* (bridge type)	50 with a span of 20 m	T-130

*Projected, test units.

between the axes of the primary and auxiliary tractors are 20 m and 12 m, respectively. The track of the load tractor is 14 m, the lifting height of the hook from the standing level is 4.06 m, and the lowering depth of the hook from the standing level is 2.5 m. The load lifting and lowering rate is 1.5 m min^{-1}. The power of the installed engines in kW (hp) is 117.5 × 2 = 235 (160 × 2 = 320). The average pressure on the soil with a nominal load of 162 kPa or 1.65 kg cm^{-2} on the hook is 88 kPa or 0.9 kg cm^{-2} in the transport position. The pipe-laying bridge-crane machine weighs 78.5 tons. The machine has been successfully tested.

Pipe-layers for laying pipe with diameters required for pipelines under construction have been and continue to be developed. Their characteristics are displayed in Table XIV.

The parameters of the mounted equipment for the pipe-layers presented in Table XIV correspond to and, in some cases, exceed the indicators of the best foreign-manufactured pipe-layers.

A fundamentally new method has been developed for laying pipeline, called the nonlifting method. The method is based on the fact that the pipeline is not only the main product of the construction process, but is also its programming element. A pipeline consisting of separate or two or three-pipe sections welded on welding platforms is assembled along the axis of the unexcavated trench after minimum route preparation. In the

Figure 11 Two-rotor excavator–pipe-burying machine.

nonlifting method, the pipes are usually insulated at the factories or on the platforms, but the insulation work can be performed during the laying process. The trench is cut under the pipeline without exertion of any forces on it. In so doing, the bottom line of the trench must correspond exactly to the lower generatrix of the pipeline, in order to make sure that it lies completely against the soil during spontaneous gradual settling of the pipeline under the force of gravity.

Thus highly accurate laying of the pipeline on the soft bottom of the trench (at any time of the year) is ensured without the use of special weight-lifting and pipe-bending machines. The pipeline is set deeper by a two-rotor excavator used for this purpose, as shown in Fig. 11.

The pipeline itself "programs" the trench line in the vertical and horizontal planes by interacting with sensors placed on the excavator and controlling the spatial position of its working organs.

During course insulation, the cleaning and insulation combine is placed on the section of the pipeline suspended above the trench and to synchronize the velocities, it is attached to the excavator.

The technology of the nonlifting method was tested while laying pipelines with diameters of 1020, 1220, and 1420 mm. When constructing pipelines in swamps or along rugged terrain, where it is difficult to use rotary excavators, single-bucket excavators set for swamp operation are used.

The new pipe-laying method, proposed by the Gubkin Moscow Petroleum Institute and SKB Gazstroĭmashina, greatly decreases the number of heavy pipe-layers engaged, since they are used only for placing and aligning pipe sections during welding. This method decreases labor costs, ensures continuity of line operation, and increases the laying rate by a factor of 1.5 or more. When laying pipe using the nonlifting method, the stress state of the pipeline is lower than for the usual method. However, the new method also has well-known disadvantages: a somewhat larger volume of soil must be excavated; in locations where the earth prism collapses under the weight of the pipeline, increased contact stresses can appear in the insulation coating, which, however, do not damage the insulation.

At the present time, when is necessary to solve the problems of more complicated and larger programs for constructing pipeline systems, new technologies are being developed for laying mainlines and more modern and powerful special construction machines are being developed to provide for complex mechanization and to increase the power available for pipeline construction.

Large-diameter gas pipelines are insulated by polymer insulating strips with mandatory polymer wrap. Polyethylene-based insulation and wrap are predominantly used. Combines that perform both the insulating and wrapping operations have been used in recent years.

Factory-insulated pipes, insulated with an epoxy coating in the case of 1020 mm pipe and with a polyethylene extrusion method in the case of 1420 mm pipe, are also used.

One of the most important indicators of the quality of insulation is considered to be the adhesion of the coating, which is to a large extent determined by good preparation of the pipeline surface (cleaning). In cleaning the pipeline in the field, just as in wrapping the insulating strips, it is very difficult, and sometimes impossible, to obtain high-quality insulation, especially under the severe climatic conditions of Western Siberia.

Factory-insulated pipe creates favorable conditions for cleaning and wrapping pipes, as well as for controlling and increasing the quality of coatings with smaller expenditures of labor and power resources.

Use of factory-insulated pipes in construction of main pipelines will increase the rate at which the insulation–laying work is performed by a factor of 1.5 (up to 3.0 km of large diameter pipeline per day). In so doing, 250–300 laborers are freed for each 1000 kilometers of large

Table XV Short List of Characteristics of Machines for Transporting Pipes and Pipe Sections

No.	Pipe carrier	Weight-lifting capacity, tons	Base machine
		Pneumatic-tyre	
1.	PV-94	9	ZIL-131
2.	PV-93	9	URAL-375E
3.	PV-204	19	KrAZ-255B
4.	PV-301A	30	MAZ-7310
5.	PTK-252	25	K-701
		Caterpillar track	
6.	PT-181	18	T-130
7.	PTG-251	25	T-330

diameter pipeline laid. Large numbers of machines and pieces of machinery are freed at the same time.

Conversion to factor insulation of pipes will not only increase the rate of mainline construction and mainline reliability, but it will also greatly decrease labor resources.

In equipping gas fields, specialized motor transport, which has special technological equipment, based on heavy truck transport, is widely used.

All of these forms of transport must have a high capcity for cross country travel and must operate reliably at air temperatures down to $-60°C$. The technological transport includes special pipe-carrying machines for delivering pipes and welded pipe sections up to 36 m long.

Pipe-carrying machines, whose characteristics are displayed in Table XV, are currently being serially manufactured.

The construction of mainlines and compressor stations in the taiga and tundra zones is characterized by low-carrying capacity of the soils during the summer and spring–fall period, as well as a deep snow cover during the winter. Technology providing high capacity for travel in difficult terrain is needed for transportation and for technological operations.

In recent years, the Tyumen's swamp transport has been developed and is serially produced by Minneftegazstroĭ plants. Its weight-lifting capacity is 36 tons (Fig. 12). The swamp transport is intended for transporting different loads in swampy and flooded areas, as well as for placing technological equipment used in mainline construction on it. The specific pressure on the soil under maximum load is $0.32\,kg\,cm^{-2}$, the engine power is 300 hp, it reaches speeds up to $12.5\,km\,h^{-1}$, and it weighs 43 tons.

Much of the hauling is done on winter roads over large distances. To

Figure 12 Tyumen swamp vehicle.

extend the range of frozen crossings, reinforcing ice columns, frozen with the help of heat pumps, are used. The distance between the rows and between the columns in a row is 6–7 m. The lower ends of the columns rest on the bottom and in the case of weak soils, on frozen foundations, which are formed by freezing the soil at the ends of the heat pumps buried in the bottom.

The columns are frozen with diameters up to 1.5 m and have high rigidity. For the distances between columns and an ice thickness of 50–60 cm, the carrying capacity of the transport strip increases by a factor of 2–2.5, which makes it possible for transport weighing up to 80 tons to pass and at the same time the stress level in the ice columns does not exceed 1.5–2 $kg\,cm^{-2}$. An algorithm was created for calculating ice crossings on a computer, with which the bending and stress at any point in the ice cover on the crossing can be determined. The maximum working stress level in the ice cover does not exceed 6.8–7.0 $kg\,cm^{-2}$.

Heat pumps are used for freezing the ice columns. These consist of hermetically sealed tubes with a diameter of 20–25 mm, including the smooth part, evaporator (5–8 m long, depending on the depth of the river), and finned part (condenser with length 0.5–0.8 m). The inner cavity of the heat pumps extend above the surface of the ice to a height of not less than 1–1.2 m. The heat pumps can be used repeatedly.

3.4. Underwater Crossings

Underwater crossings of gas pipelines are the most critical and complex parts of main pipelines.

At the present time, more than 6500 kilometers of underwater pipelines are used in water crossings. Each year, about 300 kilometers of underwater pipelines, of which 75% have diameters of 1020–1220 mm, are laid.

The cost of this program is about 100 million rubles. During the five-year period from 1981 to 1985, 1.5 times more underwater crossings will be constructed.

Pipeline diameters in underwater crossings have increased to 1220 mm. Higher work quality and reliability of underwater crossings may make it possible to eliminate reserve strings for crossings up to 250 m wide. In the Soviet Union, pipes are encased for crossings; in addition, in ammonia pipelines the space between the pipes is filled with nitrogen. A high productivity technology has been assimilated for underwater excavation: hydraulic dredges with productivity of 500 and 1000 $m^3 h^{-1}$ of soil, multi- and single-bucket dredges with excavation depths up to 20 m, as well as dredges with excavation depth up to 25 m with a productive of 250 $m^3 h^{-1}$. The pipe laying barge Suleĭman Vezirov has been successfully used to construct crossings through reservoirs and a drilling method is being developed to construct crossings under rivers.

The method of dragging the pipeline along the bottom in an underwater trench excavated beforehand with dredges is most widely used. The same method was used to construct narrow crossings for 1420 mm gas pipelines. All underwater crossings of main pipelines until recently were constructed by the special trust Soyudpodvodgazstroĭ, which has now been converted into a larger specialized All-Union organization.

The specialized organization has its own fleet: dredges, delivery tugs, barges, floating cranes, dry and liquid cargo barges, guard ships, and floating workshops. Construction of crossings of northern rivers in Western Siberia is most complicated. For example, the Vyngapur–Chelyabinsk gas pipeline crossing through the Ob' River is characterized as follows.

The river bed in the region of the crossing has a surface water width of 1250 m at low water. The depth of the water during this period reaches 14.6 m. The river flood-plain is strongly swamped, interlaced with lakes and channels, and during the high flood waters, the floodplain is flooded.

Table XVI Volume of Work for Underwater Crossing of Vyngapur–Chelyabinsk Gas Pipeline Through the Ob' River

Type of work	Unit of measurement	Volume of work	
		Riverbed	Floodplain
Underwater inspection	m^2	267 080	
Excavation and backfill of underwater trenches*	m^3	684 860 / 611 340	
Excavation and backfill of trenches* by excavators and bulldozers	m^3	469 060 / 374 090	415 920 / 345 820
Welding, insulation, and laying of pipeline	m	3 901	13 656
Ballasting of gas pipeline:			
with cast iron weights, 1.1 tons	wt	2 176	
with reinforced concrete weights	wt	7 614	
	m^3	5 263	

*Volume of soil excavated is shown in the numerator and the volume of backfill of trenches is shown in the denominator.

This rise in water level during the spring usually starts at the beginning of May and ends in August. The summer–fall low water period with stable low-water levels is very short: September–October.

The average duration of the navigational period is about 160 days.

The average thickness of the ice is 0.6–0.8 m and the maximum thickness is up to 1 m. Ice formation usually begins at the beginning of November. The flow rate at high water in the riverbed of the Ob' River attains $1.5–1.7\,m\,s^{-1}$. The shoulder on the right bank along its entire extent collapses by 3 to 4 meters per year and the left shore collapses less intensely.

The completed underwater crossing of the Vyngapur–Chelyabinsk gas pipeline through the Ob' River is one of the largest crossings in the Soviet Union. It consists of three 1.020×21.5 mm pipelines, laid in separate trenches. The excavated earth for just one pipeline exceed 1.3 million cubic meters. Table XVI shows the effort expended on the crossing.

A diagram of the pipeline crossing through the Ob' River is shown in Fig. 13.

The large underwater crossings on the Vyngapur–Chelyabinsk gas pipeline were completed within one year and, in addition, they were constructed through the Agan, Ob', B. Salym, Dem'yanka, and Turtas rivers

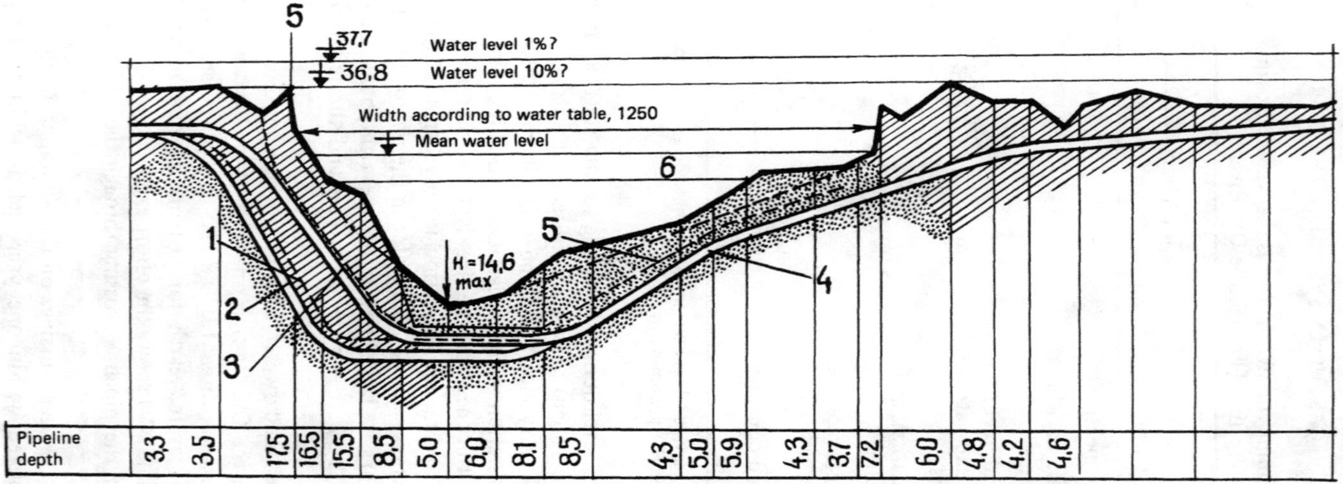

Figure 13 Diagram of the Ob' River crossing on the Vyngapur–Chelyabinsk gas pipeline. (1) Initial line of maximum erosion; (2) initial projected position of the pipeline; (3) adjusted pipeline position; (4) boundary of excavation by hydraulic dredge; (5) adjusted line of maximum erosion; (6) minimum winter level of backfill boundaries.

during the winter, while the crossings through the Tura, Pyshma, and Tavda rivers were constructed during the summer.

The Ob' crossing employed 120 men. During the preparatory period, an underwater survey was made, the approach roads and ferry crossings were inspected, and temporary and other buildings as well as the assembly and the welding–assembly platforms were constructed.

The pipes were connected on the welding–assembly platforms into three-pipe sections with the help of rotational semi-automatic welding. The sections were strung into lengths at the channel marks of the crossing by manual welding. As they were completed, the strings were hydraulically tested, cleaned, insulated, and lined, and then the weights were mounted.

The slow drop in the flood waters of Siberian rivers, decreasing the already short summer–fall period during which the water surface is free of ice, makes it necessary to construct the crossings during the winter.

As a rule, there is not enough time to complete the underwater excavation work and the underwater trenches up to the planned marks tear away during the winter after formation of the solid ice cover.

The Eniseĭ river crossing in 1969–1970 on the Messoyakh–Noril'sk gas pipeline route was the first time a gas pipeline was laid during the winter.

The 3.8 km pipeline, with a diameter of 325 mm, was laid by dragging it beneath the unbroken ice cover.

By 1979, 51 kilometers of water crossings were laid during the winter.

In the Ob' River crossing, to excavate the trenches during the two-month summer–fall period, hydroextractors, dredges, and hydraulic excavators were used. To finish off the underwater trenches, taking into account the blockage during the winter at great depths, pneumatic dredge pumps were successfully used. For pneumatic dredge pumps were connected to a stationary compressor (70 $m^3 min^{-1}$), which greatly increased the efficiency. Waterlines and shore and floodplain trenches were dug with single-bucket excavators and bulldozers. During the winter, the soil was first loosened with blast-hole charges.

Concreted pipes were used on the floodplain of the Ob' River underwater crossing. Pipes with a diameter of 1220 mm were also successfully used in this crossing, which decreased the number of pipe strings in the crossings and decreased the volume of underwater excavation work. The savings amounted to 1.8 million rubles per year.

In all crossings, the pipelines were laid using the bottom-pull method with the help of LP-1A and LP-151 winches with a pulling force of 72 and 150 tons, respectively. To decrease the negative buoyancy of the gas pipeline fitted with cast iron weights or concreted, 10-ton unloading pontoons were used.

The pulling forces used to drag the pipelines, depending on their length and negative buoyancy, varied from 20 to 160 tons.

Major problems arose in laying mechanically bent pipeline with two curves through the Lokosov channel. The pipeline, weighing 356 tons, was laid in four stages with three strings laid sequentially.

3.5. Laying Large-Diameter Gas Pipelines in Swampy Regions of Western Siberia

The lack of experience in constructing large-diameter pipelines in swampy areas in this country and abroad created great difficulties in laying the first 1420 mm pipelines in Western Siberia. New and interesting solutions of technical problems were tested on the construction of the Vyngapur–Chelyabinsk gas pipeline, which was the first pipeline of the Urengoĭ–Chelyabinsk gas pipeline system. The 1420 mm pipeline pressurized at 7.5 MPa with a length of 1547 km crossed 382 km of swamps and 334 km of flooded areas; in addition, there wre 123 river crossings. Two thousand nine hundred fifty hectares of forest were cleared along the course. Nine hundred twenty thousand tons of reinforced weights with a total weight of 480 thousand tons and 71 thousand pairs (sets) of screw anchoring systems were required to ballast the pipeline. The complexity indicator for the route, calculated according to currently used procedures, equals 3.6; the same indicator was estimated as 1.6 for the Central-Asia–Center Pipeline route and 1.8 for the pipeline from Orenburg to the western border of the USSR. Under these conditions, the gas pipeline was constructed within 18 months, while the standard time was 36 monts, and the basic linear operations were essentially completed within a single winter season.

The structural reliability of gas pipelines laid in swampy areas is primarily related to their strength and stability.

Pipe walls are in a biaxial stressed state: circumferential from internal pressure and longitudinal stresses from internal pressure, temperature differences, preliminary bending and other loads and actions, arising during construction and operation of the gas pipeline. Radial stresses are neglected due to their small magnitude.

Since the thickness of the pipe wall is determined from the condition that the metal consumption be a minimum only by the magnitude of the internal pressure, the magnitude of the limiting positive temperature differential causing additional limiting stresses must be limited. Experience in constructing pipelines in the north over a period of many years shows that the temperature differential must be taken as the difference between the maximum temperature of the pipeline when transporting gas and the temperature of the pipe as it is laid, more precisely, as pipe motion becomes restricted, which occurs approximately after backfilling ahead up to 1 km of finished pipeline. The air temperature at the time the pipeline is laid is usually used, which introduces a small error.

The temperature differential will be positive if the temperature of the gas pipeline during operation is greater than at the time the pipeline is laid. The limiting magnitude of the temperature drop is determined by an equation obtained for a state of the pipeline in which the equivalent stresses equal the working resistance of the pipe metal:

$$\Delta t = \frac{(0.25 npD_{in}/\delta) + \psi_2 R_1}{\alpha E}, \qquad (8)$$

where δ is the wall thickness (mm); n is a coefficient describing the overload of the working pressure of the pipeline chosen as 1.1 according to the norm; p is the working (standard) pressure (kg cm^{-2}); D_{in} is the inner diameter of the pipe (mm); R_1 is the working resistance of the pipe metal (kg cm^{-2}); α is the coefficient of linear expansion of the metal, equal to 1.2×10^{-5} deg^{-1}; E is the modulus of elasticity of the pipe metal, equal to 2.1×10^6 kg cm^{-2}.

The coefficient ψ_2 takes into account the two-valued biaxial stressed state based on the energy theory of strength and is defined by the expression

$$\psi_2 = \sqrt{1 - 0.75\left(\frac{\sigma_{cf}}{R_1}\right)^2} - 0.5\,\frac{\sigma_{cf}}{R_1}, \qquad (9)$$

where the circumferential stresses from internal pressure are determined from the equation:

$$\sigma_{cf} = \frac{npD_{in}}{2\delta}. \qquad (10)$$

Using the dependences presented for the Vingapur–Chelyabinsk pipeline, a limiting, positive temperature difference equal to 58°C was obtained. The transportation temperature of the gas was determined from the possibility of cooling the gas using air-cooling equipment (ACE). Starting from the computed positive temperature difference, the pipeline construction temperature, relative to different sections of the span between compressor stations, is established from the graph in Fig. 14. The distance from the compressor station along the gas flow is shown along the abscissa axis and the minimum pipeline construction temperature is shown along the coordinate axis.

An important construction parameter in laying the pipeline is the

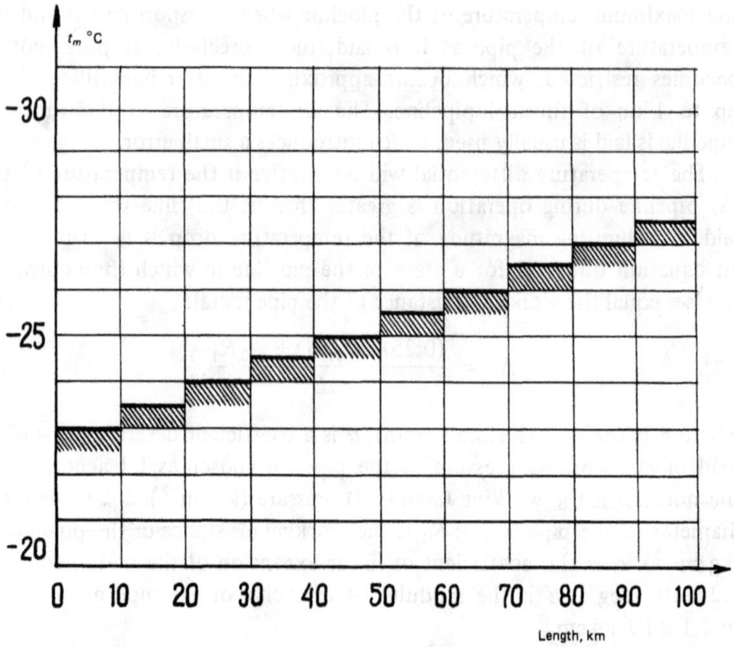

Figure 14 **Graph showing the minimum permissable air temperatures during pipe-laying and backfilling.**

minimum allowable elastic bending radius of the pipeline, which is determined primarily by the magnitude of the temperature differential, internal pressure, and physicomechanical characteristics of the pipe metal.

As the elastic bending radius decreases in laying the pipeline on flooded sections of the course and in swamps, the magnitude of the overload on the pipeline, which ensures its longitudinal stability, increases.

The minimum allowable elastic bending radius is determined from the magnitude of the stresses in the edgemost fibers of the cross section, established by standards, from the equation:

$$\rho = \frac{ED_n}{2(\psi_3(c/K_r)R_2^n + 0.15(PD_{in}/\delta) - E\alpha A t)}, \tag{11}$$

where

$$\psi_3 = \sqrt{1 - 0.75\left(\frac{\sigma_{cf}^n}{(c/K_n)R_2^n}\right)^2} - 0.5\,\frac{\sigma_{cf}^n}{(c/K_n)R_2^n}, \tag{12}$$

K_r is the coefficient of reliability, chosen as 1.1 according to the norms for key gas pipelines; R_2^n is the yield stress of the pipe metal, determined by the technical conditions on the pipe; c is a coefficient describing the category of the pipeline section, chosen according to standards; Δt is the standard temperature differential; and, σ_{cf}^{n} are the circumferential stresses from the working (standard) pressure. The results of calculations for the Vyngapur–Chelyabinsk pipeline are tabulated in Table XVII.

The increase in the diameter and working pressure of gas pipelines laid in swampy areas led to the fact that the reliability with which the pipeline functions is determined in many ways by the planned position of the pipeline and its stability in the soil. As the temperature of the pipe walls changes, the longitudinal forces increase as the square of the pipeline diameter, reaching 1700–2000 tons. At the same time, friction forces between the pipe and the soil, which counteract the pipe motion both in longitudinal and transverse directions, increase only in proportion to the pipe diameter. This situation becomes especially complicated in swamps, where the restraining ability of the soil decreases significantly.

Aside from the temperature difference and internal pressure, buoyancy affects the longitudinal stability in flooded sections. Longitudinal pipeline stability is achieved by lowering the pipeline to an appropriate depth and ballasting the line with weights or anchoring systems.

The critical longitudinal force was determined from the energy criterion for stability, including all factors affecting the pipeline. The stability of the pipeline–soil system in the presence of longitudinal compression forces was studied starting from an analysis of the total energy of the system

$$\epsilon = \tfrac{1}{2} EJ \int_0^L \left(\frac{d^2 v}{dx^2}\right)^2 dx - \tfrac{1}{2}\int_0^L S_x \left\{\left[\frac{d(v+v_0)}{dx}\right]^2 - \left(\frac{dv_0}{dx}\right)^2\right\} dx + \int_0^a Kv^2\, dx + \int_a^{L/2} (qn_p - C_p v)v\, dx, \qquad (13)$$

where S_x is longitudinal compressive force, caused by the change in temperature and internal pressure; EJ is the bending rigidity of the pipe; $v_0(x)$ is the initial bending of the pipeline; $v(x)$ is the additional displacement of the pipeline; L is the length of the buckling wave; q is the elastic part of the soil; q_{max} is the maximum restraining capacity of the soil for transverse, upward displacements of the pipe, and, K and C_p are parameters that characterize the working model of the soil.

The type of instability was determined by studying different types of instabilities.

At first, as the load increases (temperature and internal pressure), the initially curved pipeline undergoes small transverse displacements.

As the load increases further, at some point, corresponding to the displacements and buckling of the pipeline.

Table XVII Results of Calculations of Maximum Permissable Elastic Bending Radii for Different Categories of Sections on the Vingapur–Chelyabinsk Gas Pipeline

Pipeline diameter, mm	Manufacturing plant	σ_{cf} kg mm^{-2}	σ_t kg mm^{-2}	R_I, kg mm^{-2}	P, kg cm^{-2} Permissable	P, kg cm^{-2} Adopted	Δt, °C Permissable	Δt, °C Adopted	ρ, m Permissable	ρ, m Adopted
				Category 3 sections						
1420 x 16.5	Imp.	60	47	35.1	76	75	72.5	58	1250	1500
1420 x 19.5	Imp.	60	47	–	–	–	–	–	–	–
1220 x 16.8	Imp.	60	45	–	–	–	–	–	–	–
1020 x 16.0	Imp.	60	40	–	–	–	–	–	–	–
1020 x 21.5	Imp.	54	40	–	–	–	–	–	–	–
				Sections in categories 1 and 2						
1420 x 16.5	Imp.	60	47	–	–	–	–	–	–	–
1420 x 19.5	Imp.	60	47	29.1	75	75	58	58	1840	2000
1220 x 16.8	Imp.	60	45	29.1	75	75	58	58	1500	2000
1020 x 16.0	Imp.	54	40	26.3	77	75	57	57	1730	2000
1020 x 21.5	Imp.	54	40	26.3	102	75	82	57	1930	2000

Remark. Pipes with diameter 1220 × 16.8, 1020 × 21.5, and 1020 × 16 mm were used in multiline crossings through rivers.

Based on studies performed by VNIIST, working dependences were obtained that establish a relation between the acting loads and the forces (pressure and temperature differential), the elastic bending radius, and the magnitude of ballasting that ensures longitudinal pipeline stability.

To calculate an underground pipeline taking into account all actions on a computer, VNIIST and YuZhNIIGiprograz developed the complex program DOGA. This program permits calculating an underground pipeline with an arbitrary contour of the axis in the vertial plane, consisting of straight, elastically bent curves and branches, laid under different soil conditions.

The computer prints out the values of the transverse and longitudinal displacement, bending moments, and longitudinal forces along the pipeline. Then the strength calculations are performed and the deformation strength and longitudinal stability are checked. Results are printed out for all three limiting states of the pipeline. If some condition is not satisfied, then information on the location where this condition is not satisfied is printed out. In this case, it is necessary either to increase the elastic bending radius, pipeline depth, magnitude of ballasting or to make some other decision.

In type I and II swamps, concave elastic curves with a large radius can be used over the entire extent of the swamp and convex curves on the sections leaving the swamps. In so doing, a carrying capacity of 0.1–0.3 kg cm^{-2} is used for the swamp and the convex sections at the periphery of the swamp were anchored by clusters of anchors with high carrying capacity. The displacements resulting from temperature deformations on the concave arcs are directed toward the bottom of the swamp, which permits restricting the overload within the limits required to counteract the buoyancy force of the water. The restraining capacity of the return arc was not included. Clusters of anchors on the exits from swamps and on the convex sections prevented propagation of deformations arising within the swamp to adjacent sections.

Different solutions to structural problems are used to protect the startup and receiving units of the cleaning equipment, protective valves near compressor stations, and connecting strips between reserve strings on river crossings. The simplest and most expensive method is to create concrete supports, which must be able to absorb more than 1000 tons.

The stabilizing system illustrated in Fig. 15 is used successfully on norhtern pipelines. This system is installed in combined or separated startup and receiving units of the cleaning equipment. The stabilizer absorbs the longitudinal axial force that arises, and it is welded to the body of the pipe using a peta-shaped sleeve.

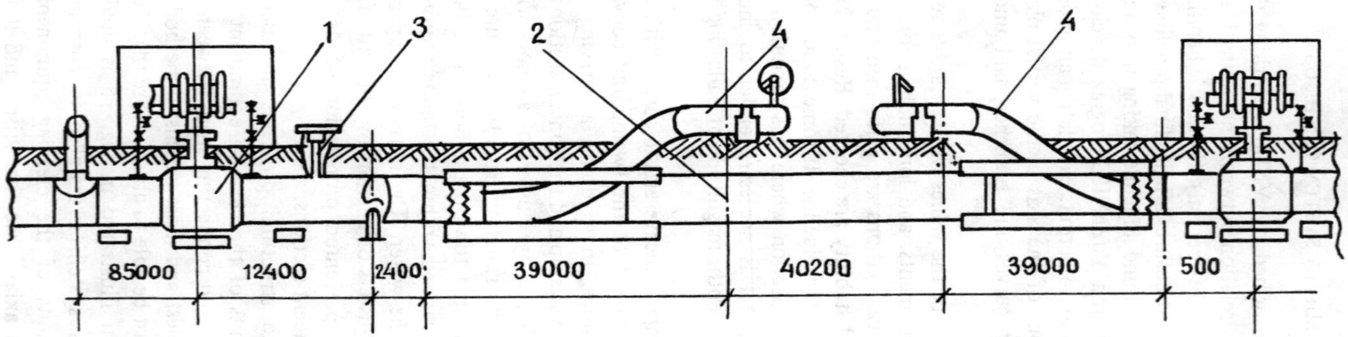

Figure 15 Stabilizer for startup-receiving unit of scrubbing installations: (1) valves; (2) stabilizing unit; (3) unit with analyzer for scrubbed gas; (4) outlet.

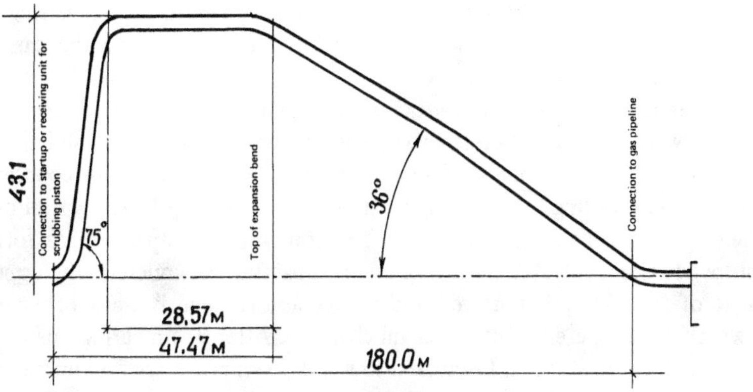

Figure 16 Diagram of expansion–support for 1420 × 19.5 mm pipeline.

The protective valves and connecting strips between the reserve strings in water crossings can be protected by installing expansion bend supports. The geometrical parameters of the configuration of expansion bend supports are determined by calculations. The choice of parameters of such expansion bends is based on the conditions limiting the stresses in the structure to magnitudes established by the standards. Figure 16 shows a diagram of the expansion bend support for a pipeline with a diameter of 1420 x 19.5 mm.

In constructing underwater crossings in the riverbed, pipelines are often ballasted by ring-shaped cast-iron and reinforced concrete overweights. To construct crossings, the engineer A. P. Kotov proposed constructing concreted pipes with a diameter of 1020 mm with monolithic ferrocement, molded on the pipe in separate sections with a length of 3.5 m. The trust Surguttruboprovodstroĭ constructed a plant for manufacturing concreted pipes.

The technological process of concretion involves the following. First, the pipes are dried and cleaned. Then a film insulation is deposited on them in two layers, which is covered by two layers of wrapping in a single pass on a special machine.

The insulated pipe is placed into the lower part of a form, consisting of three sections. Mesh reinforcement with fixatives is placed beforehand into the form to create a protective layer of concrete above the reinforcing metal. Then, in all sections, the upper part of the mesh is assembled with mandatory fastening of wooden bolster stays to ensure uniform distribution of the meash reinforcement over the perimeter of the pipe. The thickness of the protective concrete layer before the working reinforcement is not less than 30 mm, while the distance between meshes is 50 mm.

After assembling the mesh reinforcement, the upper part of the casing is installed. The distance between the surface of the pipe and the casing equals the computed thickness of the concrete.

After the upper casing is attached to the lower casing with the help of screw locks, the concrete mixture is delivered with buckets and three vibrators, mounted on the lower casing, are switched on. The concrete is chosen according to strength, resistance to frost, and density with bulk density not less than 2300 $kg\,m^{-3}$. The concrete is compacted in a form suspended on cables. After laying and compacting the concrete, the upper part of the casing is removed and the concreted pipe is transported on two cables at the ends into a steam chamber. After steam curing, the concreted pipes are stacked. Compared to well-known processes for concreting pipes, this technology permits the process to be performed rapidly and efficiently.

Concreted pipes with a diameter of 1020 × 21 mm, weighing 19 tons, and 12 m long were delivered along winter roads from the town of Surgut to the water-crossing construction site by pipe carriers.

The middle leg of the trip was 150 km and although the pipe carriers were not specially re-equipped, there was no damage to the concrete cover.

Concreted crossings with a diameter of 1020 mm were laid in the floodplains of the Ob' River through channels in constructing the Urengoĭ–Chelyabinsk gas pipeline. The underwater trench was dug from the ice with a dragline excavator. The concreted pipes were simultaneously assembled and welded at the assembly area into 36-meter sections. The sections were then laid by the bottom-pull method or directly from the ice into a lane in the ice.

3.6. *Pipeline Ballasting*

A large number of main and industrial pipelines and branches to cities and plants are laid annually in swampy and flooded areas. Approximately 6000 kilometers of pipeline requires ballasting. Thus ballasting is one of the main technological operations and often determines the rate, efficiency, and reliability of gas pipeline laying.

Ballasting is done with reinforced concrete overweights with different construction and various types of anchors. More than 3000 overweights are installed per year on pipeline routes, for which about 540 000 cubic meters of reinforced concrete and 100 000 anchors are used. Specialized crews, equipped with special technology, perform the ballasting.

Experience in constructing gas pipelines with diameters of 1220 and

1420 mm has shown that the most widely used reinforced concrete saddle weights fall off due to temperature displacements of the pipeline and due to its movement, and they do not guarantee that the pipeline will remain in the planned position.

Comparative tests and expert evaluation of weights with different construction were performed in order to choose the optimum construction of reinforced concrete weights. The weights for the 1420 mm pipelines, which were subjected to comparative tests, are shown in Fig. 17. Weights were evaluated according to 17 parameters: stability of the weight in a plane perpendicular to the axis of the pipeline, labor required to install the weights on the pipeline, complexity of manufacturing the weights, consumption of metal per cubic meter of concrete, restraint of the pipeline by the weight, shear force of the weight along the pipeline, etc.

From the results of comparative tests and expert evaluations, the best indicators were obtained for UBO weights. For this reason, these weights are presently widely used in the construction of gas pipelines. However, the search for optimal solutions of the problem of ballasting pipelines with weights continues. The installation of UBO weights on a gas pipeline is shown in Fig. 18. The weights are installed by pipelayers or crane trucks with a weight-lifting capacity of 10 tons. Protective and lining mats are placed under the collar of the weighting material on the pipeline and the collars are carefully insulated with polymer tape. Pipeline ballasting was performed by special crews.

To increase the efficiency of ballasting, and this is very important since this operation in many ways determines the rate at which the gas pipelines are laid in swampy areas, a group method is used for installing the weights. The weights were laid on the pipeline directly against one another, and their total number per kilometer of pipeline corresponded to the eequirements of the project. The group placement of weights is determined by the magnitude of buckling of the pipeline between the groups of weights. For a pipeline with a diameter of 1420 x 17.5 mm, the maximum number of weights per group is 32, while the distance between groups is 33 m.

When constructing gas pipelines during the summer in impassable swamps, the weights are installed using MI-6 and MI-10K heavy-duty helicopters. The helicopters are based as close as possible to the construction site: 300–500 meters. Sometimes, however, this distance reached 5–9 km.

The helicopter hovers above the area where the weights are stacked. The rigger, located in the dead zone beneath the helicopter, slings the overweight under the helicopter. When this operation is completed, the helicopter transports the overweight to the installation location on the

Diagram showing structure of weights	Type	Designer	Technical documentation	Tests	Use
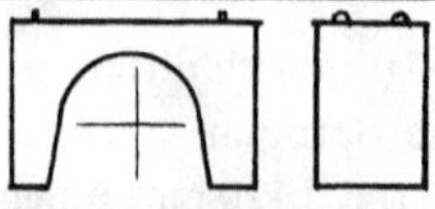	Saddle-shaped	VNIPItransgaz	Working drawings, technical specifications	Inspection and checking protocol	Yes
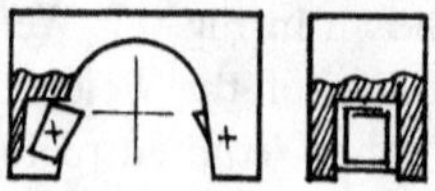	USS	VNIIST, Apsalyamov SK and M combine	Working drawings	Factory tests	No
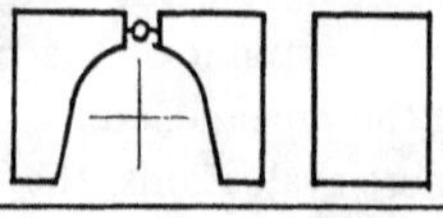	Hinged	Glavsibtruboprovodstroi, VINIIST	Working drawings	Inspection and checking protocol	Yes
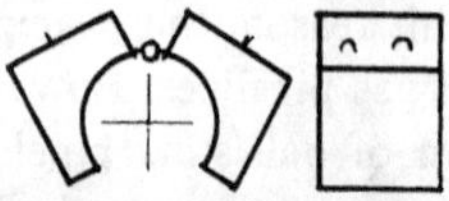	SUG	Lengazspetsstroi trust	Working drawings	Inspection and checking protocol	Yes
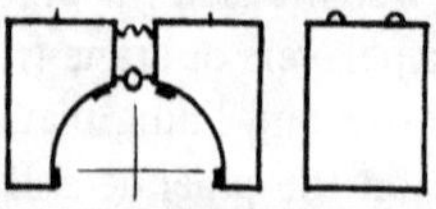	SG	VNIIST	Working drawings	Factory tests	No
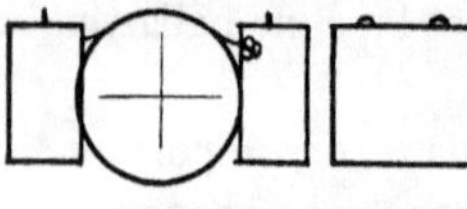	UP	EKB for reinforced concrete, VNIIST	Working drawings	Factory tests	No
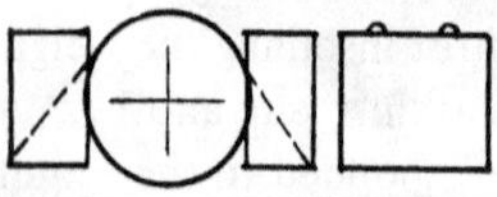	UBPshch	Shekingazstroi trust	Working drawings	Factory tests	No
	UBO	VNIIST	Working drawings, technical specifications	Inspection and checking protocol	Yes
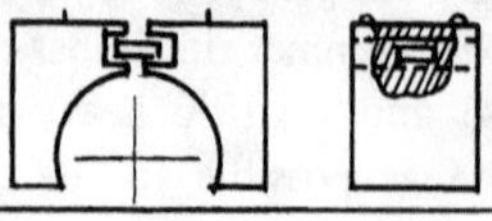	Hinged with wedge-shaped fastener	Glavneftegazpromstroimaterialy	Working drawings	–	No

Figure 17 Construction of reinforced concrete weights.

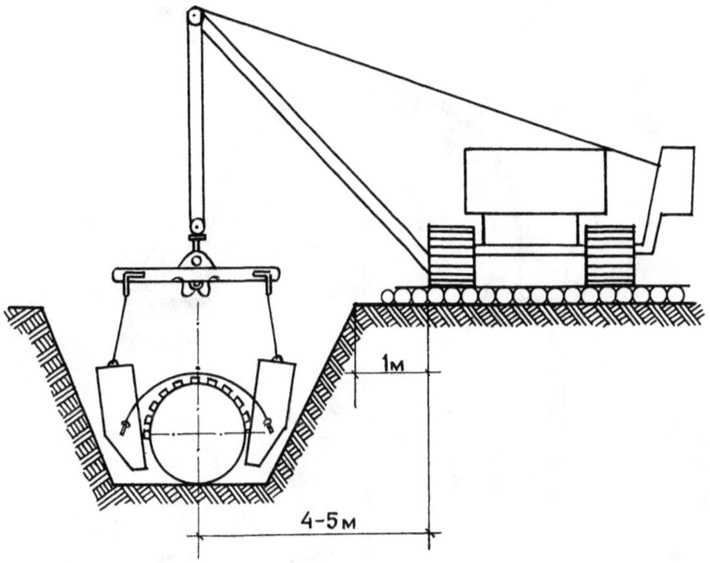

Figure 18 Mounting of UBO weights on the gas pipeline.

pipeline. The helicopter hovers above the location marked by the rigger, after which when signalled, the overweight is placed smoothly on the pipeline. Seventy weights have been installed per working day in the case of the short distance and 40 weights in the case of the long distance.

The group method has great advantages in installing overweights with helicopters.

Screw and pile anchors with different construction are widely used to ballast pipelines. Screw anchors can be used in all flooded and swampy sections of the pipeline route, as well as in swamps with peat up to 5 meters thick. The minimum depth of the screw anchors in mineral soil must be 2.4 m.

In recent years, cast iron anchors have been replaced by welded steel anchors, which have a number of advantages. There are no hidden defects in the screw blades of steel anchors. Welded anchors are simple to make, they weigh less, and they are cheaper than cast iron anchors. A smaller torque is required to turn them.

A screw anchor unit consists of two screw anchors, two tie rods, a force collar, a lining mat, and padding. Figure 19 shows a completed screw anchor unit (SAU). The tie rod has a cap and it is put on the screw part without welding. This permits transporting the tie rods separately from

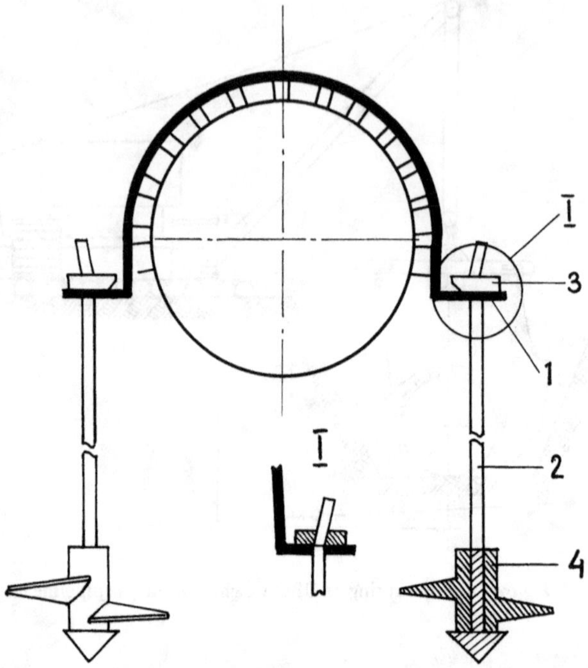

Figure 19 Screw anchoring installation: (1) force collar; (2) anchor tie rod; (3) steel plate; (4) screw part of collapsible anchor.

the screw part and increases the weight-lifting capacity of railroad platforms by 30%. Screwing is performed with the help of a bar, which is placed on the tie rod and is pushed against the base of the anchor blades with a patterned groove. The fixing joint consists of the tip of a force collar with a steel plate bent at an angle. A tie rod is passed through the opening and when a load is applied, the bent ends of the force collar are straightened out and "bite into" the tie rods. This method accelerates the anchoring process and eliminates the need to weld the collars to the tie rods. Anchors with blades having a diameter of 400 mm are usually used to secure large-diameter gas pipelines. The screw blade represents in a plate a complete loop with a definite pitch-to-diameter ratio. The length of the tie rod is determined by the depth of the anchor.

The force collar transmits forces from the pipeline to the anchor. It consists of a steel bar and it is placed on the pipeline together with lining mats and padding. The tie rods and the force collars are insulated with polymer tape, while the screw blade is insulated with a special primer and

Table XVIII Soil Carrying-Capacity Factor

Soil group	Soils	Soil carrying-capacity factor, K_s
1	Soft clays and loams, sandy loams	1
2	Sands with fine, high and average density, low moisture content, moist, and water-saturated; semi-solid, hard clays and loams	2
3	Gravelly sands, large and medium graininess, low water content, moist, and water-saturated, solid sandy loams, clays, and loams	3

bitumen-rubber mastic. The screw assemblies are manufactured at plants and all elements are delivered to the course ready-made.

Compared to concrete overweights, the anchors have a small mass and they can be easily delivered to any point on the course.

Screw anchors are buried without destroying the structure of the soil, and their weight is insignificant compared to the restraining force. The carrying capacity of the anchor assemblies is determined by the characteristics of the soil, the construction of the anchor, and the depth of the anchor blade.

The force (permissible load) on an anchor assembly B_{anc} was calculated using the following equation:

$$B_{anc} = ZK_s N_{anc} m_{anc} \tag{14}$$

where Z is the number of anchors in a single anchor assembly; K_s is the coefficient of carrying capacity of the soil, chosen from Table XVIII, N_{anc} is the maximum (critical) load on a single anchor, screwed into a group I soil to a depth of six blade diameters, taken from Table XIX; m_{anc} is the coefficient which depends on the working conditions of the anchor assembly, taken as 0.5 for $Z = 2$ and 0.4 for $Z > 2$.

Before positioning the anchors, test loadings are performed. After the maximum (critical) load is installed on the anchor, the distances between the anchoring assemblies is adjusted. The distance between the axes of the anchor assemblies (span) was determined from the equation

$$\rho \leqslant \frac{B_{anc}}{B}, \tag{15}$$

Table XIX Maximum Load on a Single Screw Anchor as a Function of the Anchor Diameter

Anchor diameter, mm	Maximum (critical) load on a single screw anchor N_{anc}, kg
100	650
150	750
200	1 350
250	2 100
300	3 000
400	5 300
500	8 300
600	12 000

where B is the working force on the anchor assembly per meter of pipeline, determined in accordance with the standards.

If it is necessary to limit the buckling (rise) of the pipeline in the span between anchoring assemblies, then the magnitude of the span is checked using the equation

$$\rho \leqslant 0.01 \sqrt[4]{\frac{384EJf}{q}}, \tag{16}$$

where E is the modulus of elasticity of the steel pipe (kg cm^{-2}); J is the moment of inertia of the transverse cross section of the pipe (cm^4); f is the permissible bending of the pipeline (cm); q is working load on the pipeline per unit length (kg cm^{-1}); the load per unit length is determined from the equation

$$q = 0.01(q_b - q_{pl} - q_{add}) \tag{17}$$

where q_b is the buoyant force of the water (kg m^{-1}); q_{pl} is the working mass of the pipeline in air (kg m^{-1}); q_{add} is the working mass of the transported gas, mass of the additional units, as well as the mass of ice that can form when gas at a negative temperature is transported (kg m^{-1}).

Data sheets, which record the carrying capacity of screw anchors in specific soils and route conditions, are drawn up for the screw anchor assemblies. To check the carrying capacity and correctness of the anchor spacing on the route, the anchors are usually pulled out every 100 meters.

The starting load for calculating the weight required to secure a 1420 mm gas pipeline against buoyant forces was taken as 1.2 tons per meter of pipe, and a starting load of 0.8 tons was used for the 1220 mm pipe.

The working carrying capacity of an anchor assembly consisting of two anchors was determined from data obtained by pulling out a single anchor,

Table XX Characteristics of Equipment for Screwing Down VAG-202 and VAG-101 Anchors

Technical characteristics	Type of equipment VAG-202	VAG-101
Torque, kg m	2000	1 000
Rate of rotation of anchor, rpm	7.5	8
Projection of boom (from edge of track), m	4.5–7.5	4.5
Rotation of boom, deg	110	90
Dimensions, mm:		
a) length	4380	7 800
b) width	5500	2 790
c) height	8000	5 000
Weight, kg	3100 (without pipe layer)	17 100

screwed down to a depth of 2.4 m. This quantity in tons divided by 1.2 tons m^{-1} and 0.8 tons m^{-1} for pipes with the corresponding diameters gave the span between anchor assemblies in meters.

Anchors with diameters up to 400 mm, inclusively, were screwed down with the help of special VAG, MBTA, MZVK, and MZA units. The VAG-101 and VAG-202 units are most widely used. The characteristics of these machines are presented in Table XX.

Figure 20 shows a MZA-205 machine for screwing down screw anchors, arranged on a TG-4 tractor. The machine weighs 17.5 tons. The working organ of the machine is a rotator, consisting of a hydraulic engine and a reducer. The maximum torque developed by the rotator is 19.6 kN-m. The maximum projection of the rod is 7 m, which permits placing anchors on both sides of the pipe. Anchors can be positioned in clusters on both sides of a pipeline using the MZA-205 machine. The cluster method of anchor placement reduces the net cost of the completed work by 5–8% and increases the productivity of machines and pieces of machinery by 20–25%. The complex of machines and equipment also includes a welding unit for welding anchors to the collar and a bitumen melting kettle for depositing the protective bitumen coating on the surface of the collars and welding collars to tie rods. The carrying capacity of the anchor assemblies is monitored by dynamometers. The organization of the anchor placement work is shown in Fig. 21.

The anchors were screwed down after the pipeline was laid in the

J

Figure 20 Machine for screwing down MAZ-205 screw anchors.

trench. During the winter, the bottom of the trench had time to freeze and it was necessary to loosen or defrost the soil in order to install the anchors. Under these conditions, it was possible to install 15–20 pairs of anchors per shift. The overall rate of pipelaying was sustained. At the present time, anchors are screwed into the bottom of the trench before laying pipelines immediately after the trench is cut. This permits installing 40–50 anchor pairs per shift.

AR-401 expanding pile anchor assemblies, which had a high carrying capacity of up to 50 tons and up to 70 tons in permafrost soils, were used for the first time in the Soviet Union in the construction of the Vingapur–Chelyabinsk gas pipeline. Figure 22 shows the construction of such an anchor system. AP-401 includes two pile anchors, a force collar, a lining mat, and padding. The pile anchor consists of a tubular rod with a diameter of 168 mm with an 8–10 mm wall and a tapered welded or cast end piece and four trapezoidal blades, connected to the rod with a hinge joint. The blades were situated in pairs on two levels with the angle of rotation in a plane between pairs equal to 90°.

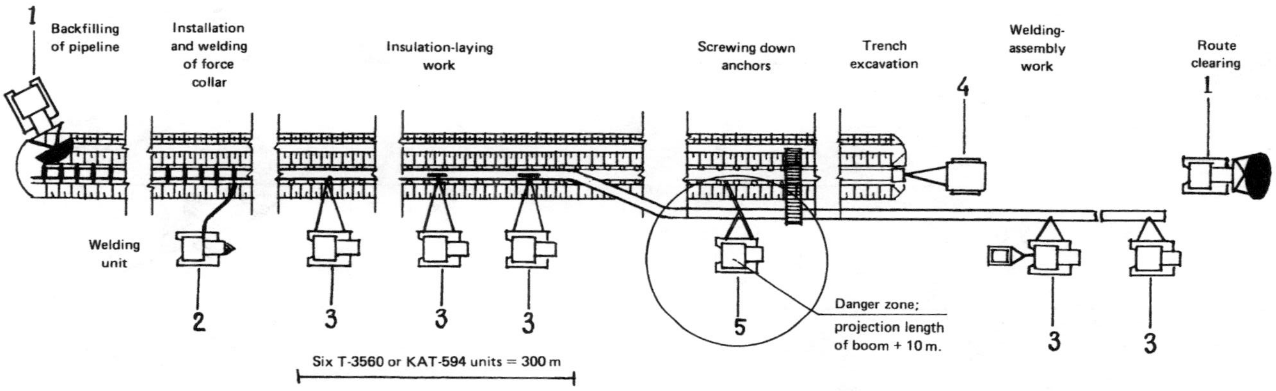

Figure 21 Organization of anchor installation work: (1) bulldozer; (2) welding units; (3) pipelayer; (4) excavator; (5) machine for screwing down anchors.

Table XXI Expansion AR Anchor

Type of anchor	Weight, kg	Average (experimental) carrying capacity, tons			Movement of anchor up to total opening of blades, m	Area of blades, m^2	Remarks
		Group 1 soil	Group 2 soil	Group 3 soil			
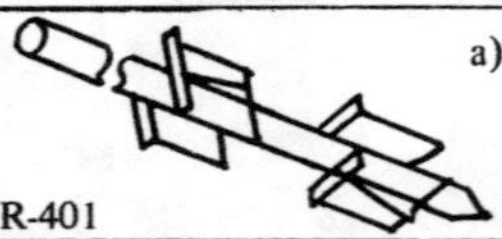 a) AR-401	442	25	42	63	1.1–1.3	1	Used for anchoring large diameter pipes in group 1 and 2 soils
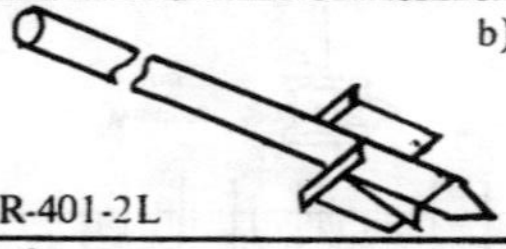 b) AR-401-2L	356	12	21	31	1.4	0.5	
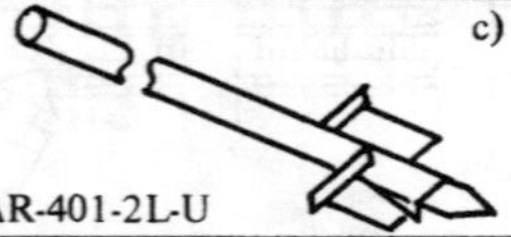 c) AR-401-2L-U	391	24	40	60	1.6	1	
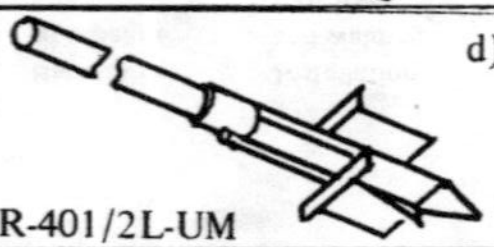 d) AR-401/2L-UM	405	20	35	52	1.6	0.98	Synchronous opening of blades
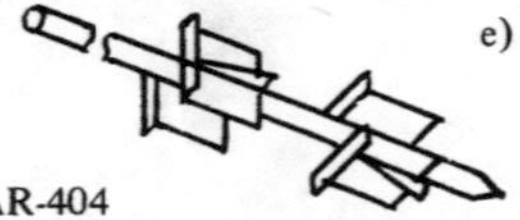 e) AR-404	145	–	–	–	–	0.5	Recommended for anchoring pipelines in permafrosts frozen into soil

 f) AR-403-D	351	21	36	54	1.6	1	
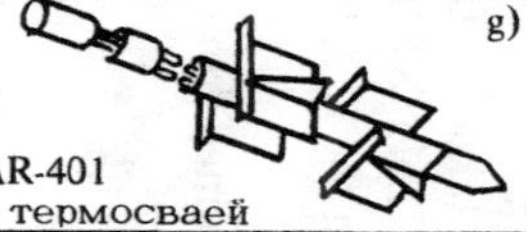 g) AR-401 с термосваей	581	25	42	63	1.1–1.3	1	Recommended for anchoring in permafrost
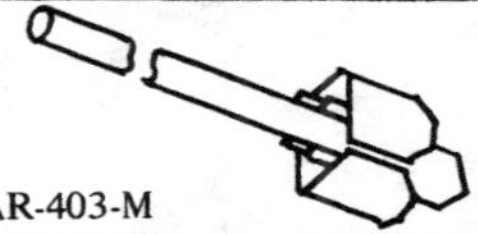 h) AR-403-M	349	22	38	57	1.7	1	
 i) AR-403-AM	324	14	24	36	0.9	0.5	
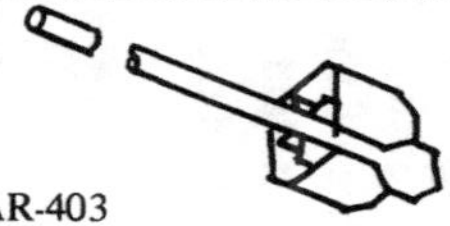 j) AR-403	125	20	35	52	1.8	1	Diameter 36 mm, steel 35; driven in with standard rod
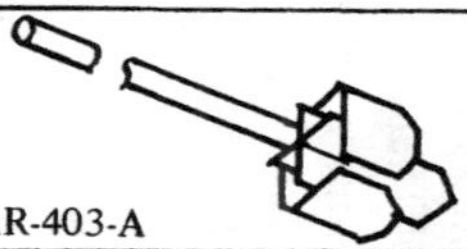 k) AR-403-A	100	12	21	31	1.0	0.5	Diameter 36 mm, steel 35; driven in with standard rod

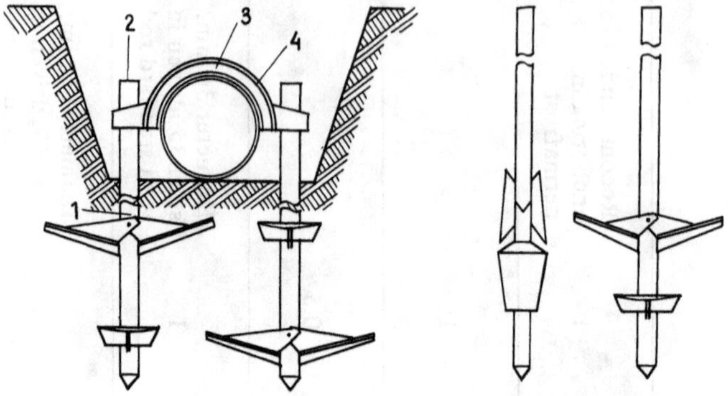

Figure 22 AR-401 anchors: (1) welded anchor; (2) force collar; (3) lining mat; (4) linear.

The pile anchor is driven into the soil and expanded by pulling the anchor up to a height of 1.2–1.5 m. Anchors are driven into soil with the help of the usual pile-driving equipment and they are pulled up into the working position by a pipelayer through the dynamometer. The opening of the blades is detected by a sharp increase in the indications of the dynamometer up to 25–40 tons, and in this case the displacement of the anchor falls in the range 110–130 cm.

The anchor unit was protected from soil corrosion by Étinol' lacquer. Table XXI presents the different constructions of AR type anchors and briefly describes their technical characteristics. The working resistance of AR type anchors with a total blade surface area $F = 1.0\,\mathrm{m}^2$ and depth $h = 5.0\,\mathrm{m}$ from the bottom of the trench can be determined in analogy to the calculation of the resistance of screw piles to pulling according to the carrying capacity for the first limiting state using the equation

$$P^1 = KmP_{lim} \tag{18}$$

where K is the coefficient of soil homogeneity, taken as 0.6; m is the coefficient of working conditions, taken from Table XXII; P_{lim} is the limiting resistance of the anchor, determined from the equation presented below or from test data.

The limiting resistance of the anchor was determined from the equation

$$P_{lim} = (AC + B\gamma h)F, \tag{19}$$

where A and B are dimensionless coefficients that depend on the angle of internal friction of the soil in the working zone, taken from Table XXIII;

Table XXII Data for Calculating Resistance of Type AR-401 Anchors with Total Blade Area $F = 1.0\,m^2$ and Anchor Depth $h = 5.0\,m$ from Trench Bottom

Soils		Working conditions factor with pulling load
Clay	a) Solid, semisolid and hard	0.7
	b) Soft	0.7
	c) Fluid-soft	0.6
Sandy	a) Low water content	0.7
	b) Wet	0.6
	c) Water-saturated	0.5
Sandy loams	a) Solid	0.7
	b) Soft	0.6
	c) Fluid	0.5

C is the specific cohesion of the soil in the working zone (tons m^{-2}); γ is the weighted mean density of the soils at a depth from the bottom of the trench to the depth marking of the anchor blade (tons m^{-3}); h is the placement depth of the blades from the bottom of the trench (m); F is the area of the blades (m^2).

Table XXIII Data for Determining the Maximum Resistance of Type Ar-401 Anchors

Magnitude of standard internal friction angle of soil	Magnitude of coefficients	
	A	B
10	6.2	2.1
12	6.6	2.4
14	7.1	2.8
16	7.7	3.2
18	8.6	3.8
20	9.6	4.5
22	11.1	5.5
24	13.5	7.0
26	16.8	9.2
28	21.2	12.3
30	26.9	17.5
32	34.4	22.5
34	44.5	31.0
36	59.6	44.0

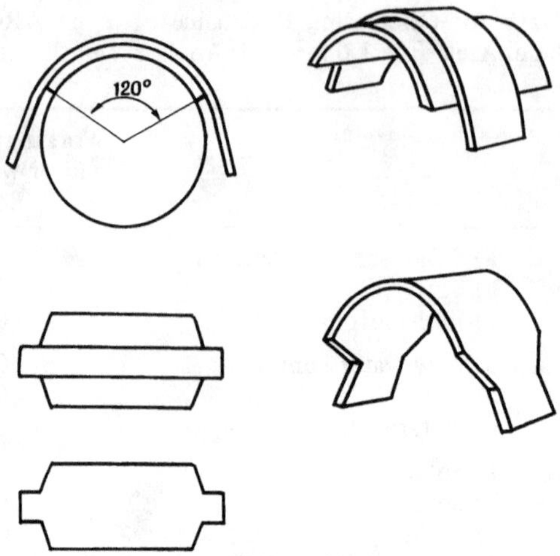

Figure 23 Diagram of force collar.

The crew for driving anchors and assembling the anchor units usually consists of 22 men. In driving anchors into frozen soil with a thickness exceeding 30 cm, a hole is drilled first. About six anchor assemblies are installed per shift.

Pile anchoring assemblies can be used in swamps, where the peat thickness does not exceed the depth of the trench. When it is necessary to secure the pipeline in a peat layer greater than 3 meters thick, the length of the bar was increased in order to lower the anchor by not less than 5 m into mineral soil beneath the peat. It is desirable to use them in locations where it is necessary to provide a large restraining force for fixed footings.

A new design for the force collar, which eliminates collapse of the pipe, has been proposed and is now in use. A diagram of the force collar is presented in Fig. 23. Application of such collars permits using maximum spacing of the anchor assemblies, taking into account the limitations on the carrying capacity of the pipe. The experimentally checked and recommended spacing for AR-401 anchor assemblies is presented in Table XXIV.

The unusually strong flood of 1979 showed that AR-401 anchors are more reliable than other forms of ballasting for securing pipelines. AR-401 anchors can be placed into permafrost soils even without cooling with heat pumps, since the frozen state of the soils is already restored within 10 days after steaming.

Comparative technical-economic indicators for different methods of

Table XXIV Experimentally Checked and Recommended Spacing of AR-401 Anchors

Diameter by wall thickness, mm	Recommended by project management, m		Calculated according to carrying capacity of anchor			Allowed by carrying capacity of pipe, m	Allowed by carrying capacity of anchor, m			Allowed by the actual carrying capacity of anchor, m			Recommended taking into account the actual carrying capacity, m		
	Soils		Soils				Soils			Soils			Soils		
	1	2	1	2	3		1	2	3	1	2	3	1	2	3
1420 × 19.5	25.8	30	23	34	46	65	23	34	46	32	46	59	32	46	59
1420 × 16.5	25	25	21	31	42	50	21	31	42	29	42	54	29	42	[50]
1220 × 16.8	30	33	31	46	62	62	31	46	[62]	44	62	81	44	[62]	[62]
1020 × 16	35	35	49	72	97*	60	59	[60]	[60]	68	97*	127*	[60]	[60]	[60]
	Recommended carrying capacity of anchor, tons		25	37	50		Experimentally verified actual carrying capacity of anchor, tons			35	50	65			

□, Limitation due to carrying capacity of pipe.
$^*L_{kp}$ = 84.8 m, the distance beyond which the pipe becomes unstable.

Table XXV Comparative Technical-Economic Indicators for Different Ballasting Methods per 1 km of 1420 mm Gas Pipeline

Indicators	Reinforced concrete weights		Anchors	
	Saddle weights	VBO	ANC-1	AR-401
Number of weights or anchors, units (complexes)	660	610	200	40
Consumption of metal, tons	17	27	20	40
Consumption of concrete, tons m^{-3}	$\frac{2650}{1180}$	$\frac{2450}{1140}$	–	–
Number of railroad flatcars for transporting weights or anchors, units	44	47	1	1
Number of truck trips, units (KRAZ-256)	330	305	3	5
Installation time for one crew, shift	6	8	4	6

ballasting 1 kilometer of gas pipeline with a diameter of 1420 mm are presented in Table XXV.

As is evident from Table XXV and from experience in using different ballasting methods, the amoutn of freight traffic can be sharply reduced by using anchor assemblies. Thus, on the Vyngapur–Chelyabinsk gas pipeline, replacement of reinforced concrete weights with anchors reduced the freight traffic by about 11 thousand rail flatcars.

As is evident from Table XXV, more than twice as much metal is used in pile anchor assemblies than in screw anchors, the estimated cost for securing 1 kilometer of 1420 mm gas pipeline with pile anchors is 31 000 rubles. The eeconomic indicators for ballasting are very important, since the scale of these operations reaches enormous proportions and often determines the rate, efficiency, and reliability of gas pipeline ocnstruction.

One screw anchor complex replaced two reinforced concrete loads on a 1420 mm gas pipeline, and one pile expanding anchor (AR-401) complex replaces 6–7 screw anchor complexes.

Figure 24 Anchors shot out of a gun.

During the Tenth Five-Year Plan, more than 400 000 screw anchor complexes were installed. In 1980, about 3000 expanding anchor complexes were installed.

Forty percent of the total length of pipelines ballasted during the Tenth Five-Year Plan were ballasted with anchors. It is proposed that 60 to 70% be ballasted with anchors during the Elevent Five-Year Plan.

An original method for securing pipelines against buoyant forces with the help of anchors fired from a gun was proposed by Glavsibtruboprovodstroĭ and is widely used. The structure of the anchors is shown in Fig. 24. A harpoon gun, placed on a tractor or other tracking vehicle (Fig. 25) is used to bury the anchor in the soil. At the present time, a series of units with a harpoon gun have been created.

The anchors are placed on both sides of the pipeline and the pipeline is secured with the help of rod or cable stays and half-collars placed on the pipeline.

A restraining capacity of the anchors of up to five tons permit securing gas pipelines with diameters up to 1020 mm using this method. The restraining force of anchors fired from a gun can be increased.

Figure 25 Harpoon gun, used for burying anchors in soil.

3.7. Construction of Compressor Stations

Construction of compressor stations on gas pipelines is distinguished by prodiguous problems, characteristic of this type of construction. The stations are located along the pipeline route according to a hydraulic calculation and, more often than not, the stations end up on sites that are far away from cities, industrial centers, and roads. This is especially true for Western Siberia, where compressor stations are constructed in regions with severe natural and climatic conditions, on sites with difficult access spread out over an enormous territory, on permaforst, and in swamps.

The specific problems involved in the construction of compressor stations involve the following: low unit cost per site combined with high labor costs, large number of different types of structures on the construction site, diverse forms of construction work, and dispersion of construction sites over enormous areas of the country. The relative number of compressor stations constructed in distant, unpopulated regions of the country is increasing continuously (60% of the compressor stations are located at distances of 25–100 km from populated points and from railroads; for 10% of compressor stations, these distances are even greater: 200–1000 km), which makes it very difficult to attract a labor force and to provide it with housing facilities, food, and equipment.

Modern compressor stations are large and complex installations, which include a large amount of technological and power equipment, control and automatic systems with installed power up to 80 000 kW, and cost up to 20–30 million rubles. It is very difficult to provide the required, vital support, including construction of setllements for the service personnel.

The standard periods of time alloted for construction of compressor stations are very short with only minimum amounts of time available for site development. All construction operations are performed under conditions of complete autonomy of the construction site.

For this reason, industrial construction techniques, advanced and technical solutions of construction problems (aluminum panels with efficient insulation, surface laying of pipelines, etc.), continuous-line combined methods for performing construction and assembly work, fully-equipped block method, and a high degree of mechanization of work are used in constructing compressor stations. Construction of compressor stations in the North presented the greatest difficulties. For this reason, in this section, we shall concentrate on the construction of compressor stations in Western Siberia (Fig. 26). During the past ten years, seven pipeline systems delivering gas to the center of the country, Urals, and Western and Eastern Siberia have been constructed.

Every year, six or seven compressor stations were constructed in Western Siberia. In the near future, 10–12 compressor stations will have to be constructed here. In so doing, it is especially important to decrease the time required to construct the compressor stations so as to match their construction rates with those of the linear parts of the pipeline and to put the entire pipeline into operation simultaneously at the planned operational capacity.

Construction of compressor stations in Western Siberia began in 1970. During the ten years from 1970 to 1980, 418 gas-pumping units with

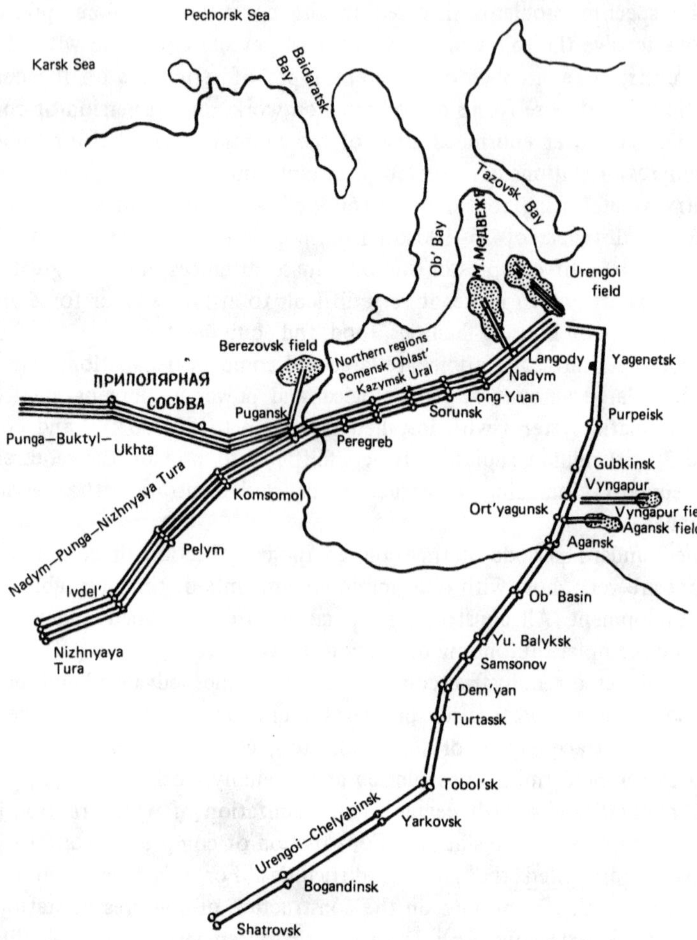

Figure 26 Diagram of pipelines with compression stations in Western Siberia.

powers of 6, 10, 16, and 25 thousand kW (GTN-6, GTK-10, TGK-16, GTA-25I) and 8STD-12.5 electric drive units were installed in compressor stations.

The complexity of and time required for construction of compressor stations are in many ways determined by the gas-pumping units used. The largest number ofunits installed (about 40%) were GTK-10 units.

Analysis of data on the time required to construct compressor stations with approximately the same capacity with different units in the North

Table XXVI Construction Time for Compressor Stations with Different Aggregates

Type of aggregate	Total installed capacity, thousand kW	Total construction time, months	Construction time scaled to one month per thousand kW
STD-12.5	900	144	0.16
Coberro	420	105	0.25
GTK-16*	96	24	0.25*
GTK-10	1850	477	0.258
GTA-25I*	150	46	0.307*
GTH-6	685	357	0.521

*The data for compressor stations with capacities of 16 and 25 000 kW aggregates are not typical, since they were obtained while constructing the first stations with such aggregates. The total and unit construction times can be expected to decrease per 1000 kW for compressor stations with high capacity aggregates.

showed that the average time for a single compressor station was 20.3 months for stations with GTK-10 units and about 16 months with 8STD-12.5 units. The construction time was shorter than the standard time. The Dem'yan compressor station with GTK-10 units on the Urengoĭ–Chelyabinsk pipeline was constructed in nine months instead of the standard 23 months. The construction time of a single compressor station is steadily decreasing.

The conditions under which compressor stations are built and the solutions of structural and equipment problems differ for compressor stations with different units. For this reason, a more accurate comparison of the construction time can be made by determining the construction times per 1000 kW of installed capacity (Table XXVI).

Compressor stations with STD-12.5 electric drive units have the shortest construction times, which is completely understandable since the amount of construction work required for compressor stations with electric drive is greatly reduced.

The total actual cost of compressor stations constructed in Western Siberia in 1970–1980 was 769.5 million rubles. In addition, the volume of construction and assembly work was twice as great in 9180 as in 1970.

The actual cost of 1000 kW of installed capacity was 0.43 million rubles with six GTN-6 units, 0.32 million rubles with six GTK-10 units, and 0.20 million rubles with eight GTK-10 units. The construction costs were lowest for compressor stations with electric drive (STD-12.5), whose actual cost was 0.09 million rubles per 1000 kW of installed capacity. The cost of a compressor station depends to a large extent on how isolated

Table XXVII Distribution of Labor in Constructing the Purgeiskoi Compressor Station on the Urengoĭ–Chelyabinsk Gas Pipeline

No.	Name of equipment	Labor, man-days
1.	Compressor plant with 5 machines	17 000
2.	Compressor plant with three machines	9 850
3.	Service-auxiliary anex	4 050
4.	"High" side of compressor station	5 800
5.	"Low" side of compressor station	5 100
	Total for compressor stations	41 800
6.	Area for air-cooling equipment	2 400
7.	Dust scrubbing area	900
8.	Connecting pipelines (loops)	4 000
9.	Processing pipelines between plants	4 000
10.	Block-containers and SKZ	3 000
11.	Engineering equipment and connections	3 700
12.	Miscellaneous building and structures	5 500
13.	Preparatory period (temporary buildings and structures	2 500
14.	Organization and intrasite roads	4 000
15	Entrance roads and careers	2 700
16.	Startup-adjustment operations	1 000
	Total:	74 500

it is, on the complexity of transportation schemes, and on the methods for supplying equipment, materials, and structures.

The plans for compressor stations, as well as the technology management used in their construction, are constantly being improved. In 1975, 160–200 thousand man-days of labor with a monthly average labor force of 400–420 men were required to construct a compressor station.

The introduction of the fully-equipped block method for constructing compressor stations decreased the labor costs by 20%. The labor involved in constructing compressor stations with GTK-10 and STD-12.5 units was 140 and 100 thousand man-days, respectively, with a montly average work force of 300–350 men. The continuous-line combined method for performing the construction and assembly work decreased labor costs by 30%. The labor involved in constructing the Purgeĭ compressor station on the Urengoĭ–Chelyabinsk gas pipeline constituted 74.5 thousand man-days with a monthly average work force of 300–310 men. The distribution of labor required to construct compressor stations is shown in Table XXVII.

Plans have now been developed for compressor stations witha modular layout of the main technological equipment (GPATs-6.3, GPU-10, and

GTK-16), a concentrated arrangement of the general construction plan, and unified auxiliary technical equipment completely outfitted at the factory, including combined power generating block (PGB) and a power service and repair block (PSRB), arranged as cellular block-box units (Fig. 27). The unit connecting the compressor station to the gas scrubbing and cooling installations is now closer. As a result of such modernization, the required labor has been reduced by one-half and now constitutes 30–35 thousand man-days per compressor station. In addition, it may be expected that the construction time will decrease by 2–2.5 months.

The effect of all organizational and construction–technological improvements on the construction efficiency of compressor stations is shown in Fig. 28. Analysis of the graphs shows that the average monthly labor required in the first three methods of constructing compressor stations differ little from one another. Moreover, the number of workers employed increases somewhat in the continuous-line combined method and with comprehensive use of advanced techniques, which can be explained by the more extensive assembly work and work requiring skilled labor, as well as by the decrease in the total construction time. With the use of the modular method of constructing compressor stations, the work force decreases by 18–20%.

The planning institutes end the project documentation with the section "Construction Organization Plan" (COP), while the construction organizations, either with their own efforts or with the help of the project technological institute Orgneftegazstroǐ or the Orgtekhstroǐ trusts, which are part of the main construction management, put together the "Construction Schedule" (CS). These plans comprise the technical documentation for management and scheduling of the construction work. It includes the transportation plan, the general construction plan, erection of temporary buildings and structures at the construction base, servicing of motorized transport, various types of storage facilities, geodetic subdivision, construction timetable, system construction charts, as well as charts for delivery of equipment, structures, materials, machines, and machinery to the construction site, and movement of the labor force. A standard plan of a temporary camp is developed and standard technological charts are used for constructing separate structures or for performing various types of work.

The institute Orgneftegazstroǐ and the Orgtekhstroǐ trusts use the best modern management methods and construction technology in their work execution plans, they supervise the execution of their plans, and they generalize experience in construction, which permits improving the technological solutions of construction problems in subsequent work

Figure 27 Compressor station built in block form.

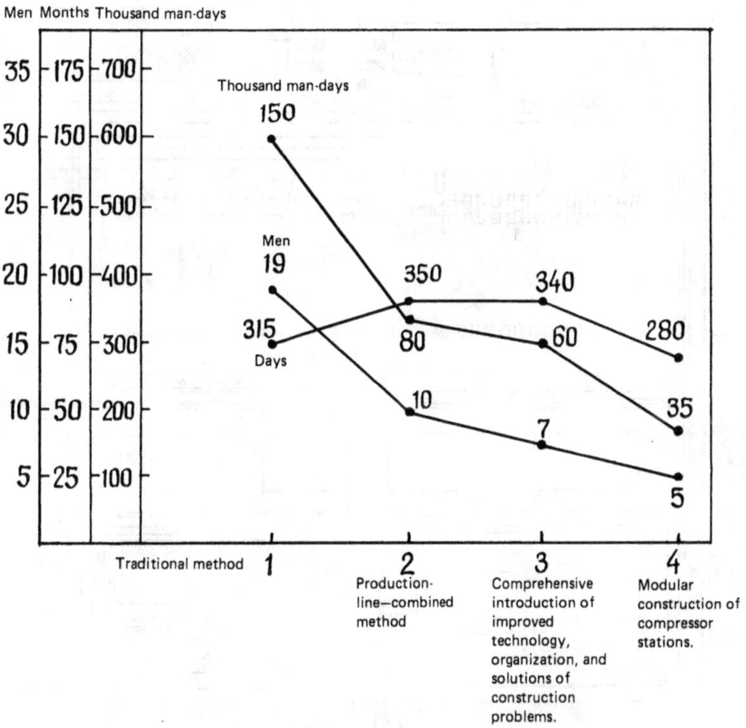

Figure 28 Dependence of construction time, labor costs, and number of workers on the method used to construct compressor stations.

plans. Figure 29 shows the general construction plan for a compressor station with GTK-10 units. Experts estimate the for the period 1976–1980, productivity of labor increased on the average by 1% due to the use of the work execution plans.

Compressor stations are, as a rule, constructed by the expeditionary–work-shift method. The workers live in special mobile camps. The work-shift camps are constructed according to standard plans, using mobile containerized blocks for housing with sanitary and domestic facilities, clinics, and recreational buildings. The camps are constructed during the preparatory period of compressor station construction. This period, which constitutes 12–15% of the total standard duration of construction of a compressor station, includes: site preparation (surveying, clearing of forests, stump removal, landfill in swampy regions, and vertical grading); construction of temporary administration and accounting buildings,

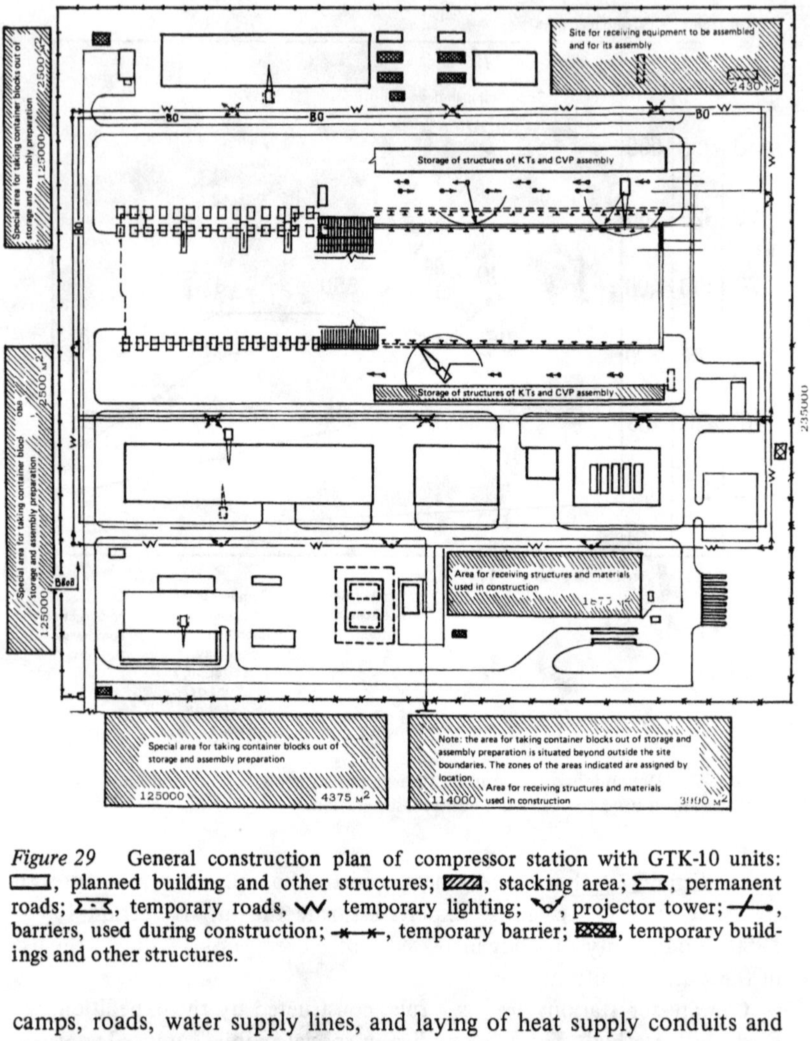

Figure 29 General construction plan of compressor station with GTK-10 units: , planned building and other structures; , stacking area; , permanent roads; , temporary roads, , temporary lighting; , projector tower; , barriers, used during construction; , temporary barrier; , temporary buildings and other structures.

camps, roads, water supply lines, and laying of heat supply conduits and power lines; and, erection of permanent buildings and structures, required for construction.

The temporary construction base is designed using standard portable structures.

The construction base, for example, for the Bogandin compressor station on the Urengoĭ–Chelyabinsk gas pipeline, was located at a distance of 500 m from the compressor base and included the following:

a concrete mixing installation with a productivity of 15 $m^3\,h^{-1}$;

two water reservoirs with a capacity of 25 m^3;

a store of fuel and lubricants consisting of four stockpiles, 25 m^3 each;

a temporary watermain, which was part of the permanent water main;

three artesian wells, comprising the permanent water supply, drilled 200 m from the construction site;

heat supply for the construction base and the camp from a permanent VVD-1.8 boiler;

before installation of a permanent power line (LÉP-10 kV), three mobile generators (PÉS-200), two working generators and one reserve generator;

during the winter, a 25 kW air heater for pre-startup heating of 42 automobiles.

The technical equipment, construction buildings and materials, and containerized blocks are delivered to the construction site by water, railroad, and motor transport. The average freight volume for a single compressor station is 75–80 thousand tons.

The average distances from the factories and suppliers and storage depots to the compressor station sites in Western Siberia are presented in Table XXVIII.

The transportation schemes are distinguished by their great complexity, there are no year-round roads (intensive use of winter roads is possible for only 3–4 months of the year), and loads must be accumulated during the year and tranported en masse (up to 70%) by motor transport within a short period of time. This makes it difficult to organize and manage the construction of compressor stations in the North.

The specific regional conditions for supplying gas works and constructing compressor stations require designs in industrial construction

Table XXVIII Average Distances from Factories–Suppliers and Storage Depots to Compressor Station Sites on the Nadym–Punga–Ukhta and Urengoĭ–Chelyabinsk Gas Pipelines

No.	Type of transport	Compressor station for Nadym–Punga–Ukhta gas pipeline, km	Compressor station for Urengoĭ–Chelyabinsk gas pipeline, km
1.	Railroad	1200	800
2.	Waterways	2200	3700
3.	Motor transport	350	250

with minimum mass, since the scheme for delivering loads can be extremely complex and it is often necessary to use helicopters.

The search for fundamentally new solutions to construction problems ended with the development of the advanced method of fully equipped block construction. The fully equipped block method transforms the construction process into an assembly process. Most of the work is transferred to specialized factories and assembly plants, which, rather than delivering hundreds of parts, deliver directly to the construction site complete modular construction blocks with technical equipment, control and measuring instruments and automatic equipment, blocks of operating equipment to be placed outside buildings together with the connections, and pipe and other assembly units.

This method has been completely realized in the Tyumen region, where a fundamentally new base has been developed in the form of a complex of specialized plants and assembly plants, a system for interconnecting the flows of resources has been developed, the problems of transporting blocks by water, railroad, and air transport as well as by all-terrain vehicles, complete site preparation for year-round construction, and introduction of production-line construction–assembly operations have been solved.

The regional system for realizing the fully-equipped block method in Western Siberia includes:

a scientific-research and planning institute, which looks for ways to improve the method, the design of separate containerized blocks, as well as fully-equipped block units;

unification of Sibkomplektmontazh with factories and assembly plants, which prepare and completely equip the block installations, units, and structures, as well as mobile transport–equipment columns, which deliver assemblies and structural subunits, completing their assembly;

general contractor construction trusts, stationed in regions of Western Siberia, specializing in preparation of and engineering support for construction sites.

This is what constitutes the construction "assembly line" in the fully-equipped block method of construction.

The primary construction teams perform the entire complex of assembly work with no limitations on their range of operation. These teams constitute large, fully-equipped brigades, consisting of workers specializing in several professions.

The best brigades perform up to 3–6 million rubles worth of construction–assembly work per year.

Further improvements in and increased scales of fully-equipped block construction of compressor stations in Western Siberia will further decrease construction times and labor costs, primarily at the construction site, where the labor costs are very high.

At the present time, all auxiliary construction in compressor stations is performed using the block method. Gas pumping units are placed in special enclosures or in prefabricated buildings.

Construction of various units on the survey datum, increased unit capacity, miniaturization of technological equipment, and placement of all communication lines above ground will further increase the impact of the fully-equipped block method of constructing compressor stations. Extensive use of this method required improvements in the entire investment process and transferring all planning, acquisition, construction, and equipment setup operations to the construction industry, i.e., a transition to a new form of construction: "turn-key construction".

The higher rates of construction and the seasonal nature of pipeline construction work make it necessary for construction organizations to be mobile.

Mobility can be achieved if these organizations have highly qualified, experienced personnel, prepared to relocate to a different region when necessary.

In 1974, 116 million rubles of construction and assembly work was performed using the fully-equipped block method, including 100 million rubles in Western Siberia. This constitutes 28% of all surface construction for the industry as a whole and 43% of all construction in Western Siberia.

In 1980, 620 million rubles or 61% of all surface construction was performed using the fully-equipped block method; for Western Siberia the figure is 450 million rubles or 66%.

The Ministry presently produces containers for living, recreational, and industrial facilities, comprising sections of prefabricated and fully-equipped buildings, with a total usable area exceeding 0.5 million square meters per year.

For the industry as a whole, due to the introduction of the fully-equipped block method, the productivity of labor in constructing surface structures was 25% higher in 1980 than in 1976, overtaking the average rate of growth of production for the Ministry. The unit (taking into account the growth in the drive power) construction time for compressor stations during the period 1971–1980 decreased by 40%. These results are all the more important in that they were achieved for higher relative amounts of construction–assembly work, performed under unfavorable conditions in the North and in Western Siberia, and higher average unit capacity of compressor and pumping stations constructed.

The process of decreasing the construction time and increasing the total number of compressor stations put into operation is also combined with a considerable (30–50%) decrease in labor costs, especially at the construction site.

We should also point out a very important result of the introduction of this method: lower, by 1 million rubles, volumes of construction–assembly work and fewer workers in the traditional professions, characterized by low output and high relative labor input.

This, by no means complete list of results due to the introduction of the fully-equipped block method, is a consequence of a number of organizational–technical actions taken during these years. These actions include, first of all, the creation of assembly–outfitting plants. Such plants have been created and continue to be developed in Tyumen, Ukhta, Shchelkov, Ufa, Al'met'evsk, Izhevsk, Volgograd, the Ukraine, Orenburg, and Belousovo.

The impact of the introduction of the fully-equipped block method on the national economy, neglecting the effect on the consumption of oil and gas, according to data compiled by the Scientific-Research Institute of Economics of the USSR State Construction Office, amounts to 418 thousand rubles er 1 million rubles of capital investment; the impact is an order of magnitude higher if the effect on consumption is included.

The Eleventh Five-Year Plan calls for the following volumes of construction using the fully equipped block method, in millions of rubles:

1981	1982	1983	1984	1985
776	932	1088	1244	1415

In 1985, all surface objects in the gas and oil industries will be constructed using this method.

Further increase in the efficiency of the fully-equipped block method requires accelerated development of new high-efficiency block equipment for surface construction in the oil and gas industries (gas pumping units, equipment for use in oil and gas fields, etc.).

The relative capital investment in the development of gas-transportation systems constitutes 70–75% of the total cost of developing the capacity for producing and transporting gas in the country as a whole.

Delay in bringing compressor stations on line sharply decreases the planned productivity of main gas pipelines. The use of the fully-equipped block method permits matching the construction times required for the linear part of the pipeline and for the compressor stations.

New, explosion-proof electric drive units without basements with capacities of 4 and 10 thousand kW, as well as gas-turbine GPA with capacities of 6, 10, 16, and 25 thousand kW, have been developed in recent years.

The prospects for using stationary gas-turbine installations with capacities of 10, 16, and 25 thousand kW and light-industrial units, based on aviation and marine units, have been determined. The relative contribution of light industrial units in the total balance in the future will exceed 50%. The use of such units permits using different solutions of problems arising in construction of compressor stations. At the present time, a single 10 000 kW unit requires 2800 m^3 of enclosed space, while a new 25 000 kW unit requires only 1800 m^3. A compressor station with a capacity of 75–80 thousand kW based on 10 000 kW units will require 8–9 machines, but a compressor station based on 25 000 kW units will require only 3–4 machines.

The new 16 and 25 thousand kW units have boosters with complete compression (full-head), i.e., it will not be necessary to construct compressor stations with two-step compression. Moreover, the collector scheme for connecting such units will simplify assembly and testing of the production pipelines.

The time required to construct compressor stations with such units will decrease and the cost of the construction and assembly work will decrease by 15 to 40% depending on the type of unit.

The new units based on marine turbines, whose technological level and suitability for industrial construction surpass the units manufactured by many foreign companies, are of greatest interest.

Assimilation of northern pipeline routes will be based primarily on the use of light industrial turbines. Aside from industrialization of the construction, the use of such units provides significant conveniences for the operating personnel because these units can be easily maintained due to the fact that the power plant can be rapidly removed and repaired at the factory. Here lies their advantage over stationary units with capacities of 16 and 25 thousand kW, which are more suitable for use closer to the center of the country.

A new dust trap with a capacity of 20 million cubic meters of gas, an air cooling apparatus with a cooling capacity of 5 million kcal h^{-1}, and other equipment have recently been developed.

4. Layout of Gas Fields

The layout of gas fields, especially under the conditions in Western Siberia, located in permafrost and swampy regions presents great technical and organizational difficulties. A considerable fraction of the invested capital is expended on gas production and preparation of gas for long-distance transportation. Short periods of time are allotted for assimilation of fields. Thus the periods allotted for assimilation of the first-class Urengoĭ field are one-half the periods allotted for the Medvezh'e field.

Large fields are laid out using wells with large diameters (not less than 150 mm as measured with respect to the lift column). The productivity of such wells, the working production rate for the entire period of operation, is about 0.8 million cubic meters of gas per day and can reach 1.5 million cubic meters and more per day.

As a rule, a clustered distribution of well in blocks of 6–8 wells per cluster is used. This permits changing over from a looped layout to a mainline layout, in which a main with a diameter of 400 mm and productive capacity of 6–8 million cubic meters of gas emanates from each cluster.

The total capacity of installations for complete preparation of gas has been enlarged. At first, installations for complete preparation of gas (ICPG) with a capacity of 3 billion cubic meters per year were built; recently, ICPG with a capacity of 10–15 billion cubic meters per year have been built. This was made possible by the development of production lies and fully-equipped block installation with a capacity of 5 million cubic meters per day. A large job involving the creation of block structures at the level of machine-building parts was executed. The Minkhimmash and Mingazprom plants are currently performing a large volume of technological installation work using the fully-equipped block method.

Fully-equipped block layout of gas fields required that the solutions of technological and engineering problems be unified first. Two basic schemes were proposed: low-temperature separation with injection of glycol and with the use of turbo-expansion engines and absorption schemes. These two schemes formed the foundation for the development of a standard fully-equipped block layout. A series of layouts were unified; a production capacity of 1, 3, and 5 million cubic meters per day with input pressure 16.0 MPa and output pressure 8.0 and 6.4 MPa was adopted. The temperature fluctuated from 0 to 100° C, while the condensate content varied from 0 to 100 g cm^{-3}.

Based on the unified series, fully-outfitted block equipment for low-temperature separation of gas with a capacity of 3 and 5 million cubic meters per day and output pressure of 7.5 MPa with injection of glycol and turbo-expansion engines was developed and introduced into the industry. Equipment with a production capacity of 10 million cubic meters per day has been developed.

Single-function and multifunction aggregate block structures were used as a basis for developing fully-equipped block installations. A single-function structure essentially includes technological apparatus and the aggregate structure includes a series of apparatus unified in a single housing with the required connections, which permits execution of the complete production cycle in the aggregate. Block compactness is achieved by using equipment that is capable of performing the production process under high intensification, as well as optimal layout of equipment and connecting pipelines. The cabinet arrangement for heating the apparatus and fixtures is replaced by heatsatellites.

Special attention is directed toward arranging blocks on the site and establishing pipe connections between blocks in order to simplify joining and interblock communication. The number of fixtures between the gas lines and the different apparatus and blocks was decreased by 30%. The blocks are equipped with the required control equipment and control and measuring instruments and automatic equipment.

The 150 modifications of blocks developed using standard technological schemes permit laying out the most diverse gas and gas-condensate fields.

An example of an aggregated block is a new type of absorber. In the past, gas preparation at the Medvezh'e field required three block installations (preliminary scrubbing separator, absorber, and filter-separator). With aggregation, all three blocks are now enclosed in a single apparatus and an aggregate for absorption drying of gas under a single housing has been developed. In so doing, not only was a new arrangement created, but the process was intensified as well. New contact separation trays were used, as a result of which this apparatus with a diameter of 1200 mm already transmits 5 million cubic meters per day. In the past such an apparatus weighed 80 tons, but now it weighs about 29 tons. Both the area and construction costs have been decreased. Such aggregation is most promising for realizing solutions of technical problems arising with accelerated assimilation of gas fields. The next stage in this work was to develop standard modules for installing low-temperature separation equipment and equipment for absorption drying of gas. Ten standard sizes and categories of fully-equipped automated modules for gas processing installations in gas fields have been developed. They include modules for storing gas and for primary separation, low-temperature separation with a capacity of 5 million cubic meters per day at a pressure of 160–80 MPa and 3 million cubic meters per day at a pressure of 100–64 MPa, absorption drying of gas with a capacity of 5 million cubic meters per day, and gas gaging and debutanization of the condensate.

The development of prefabricated modules will greatly increase the level of industrialization in outfitting gas fields and will accelerate the rate at which the equipment is put into operation since each module includes, together with fully-outfitted block equipment, interblock pipelines, cable fittings, fittings for interblock pipelines, control and measuring instrumentation and automatic equipment, and a system for automatic module control.

The development of the modular fully-outfitted block equipment does not, however, solve all problems arising in laying out gas fields. The modules only provide the auxiliary equipment. Analysis of the structure of capital investmetn in ICPG-2 at the Medvezh'e field shows that buildings and structures constitute 83.1% of the total cost of the gas preparation equipment or 21.9 million rubles. With an actual cost of ICPG-2 of 29.8 million rubles, the production structure without the equipment costs 11 million rubles. This fact indicates that the use of the fully-equipped block method of construction with equipment delivered in modules decreases the capital investment per ICPG by a minimum of 10 million rubles.

If the entire ICPG production complex is included in the fully-equipped block method, then the capital investment can be reduced by a factor of two.

The comparative technical–economic indicators of the traditiona variant of ICPG and using the fully-equipped block method for ICPG-2 at the Medvezh'e field are displayed in Table XXIX.

To achieve maximum efficiency in laying out the Medvezh'e gas field, the following were used:

primarily fully-equipped block structures;

prefabricated technological equipment in block form;

bitumen–perlite thermal insulation, which eliminated the need to lay pipelines in a trough;

AN-22 aircraft with a weight-lifting capacity of 80 tons to deliver loads from Tyumen to the Medvezh'e field.

As a result of the use of new techical and organizational methods, the ICPG-2 at the Medvezh'e field was constructed in 4.5 months instead of the standard 26 months.

The numerous water ways in the regions with gas fields make it possible to float large blocks, assembled in assembly–outfitting plants, to the construction site. Specially designed fully-equipped blocks weighing 400 tons were developed for this purpose. Sibkomplektmontazh has

Table XXIX Comparative Technical-Economical Indicators for the Traditional Variant of ICPG and Fully-Equipped Block for the ICPG-2 in the Medvezh'e Field

No.	Indicators	ICPG variant		Ratio of indicators
		Traditional	Fully-equipped-block	
1.	Productivity, billion $m^3 y^{-1}$	10.0	10.0	1.0
2.	Capital investment, million rubles	29.8	10.2	2.9
3.	Unit capital investment, rubles/1000 m^3	2.9	1.0	2.9
4.	Cost of technological housing with equipment, million rubles	16.3	5.0	3.2
5.	Cost of auxiliary constructions and other type of work, million rubles	13.5	5.2	2.6
6.	Area of general plan, ha	6.7	1.3	5.1
7.	Development index	0.3	0.4	1.1
8.	Volume of building with technological housing, gas gauging, loop inlet unit, auxiliary building block, 1000 m^3	60	10.8	5.5
9.	Unit weight of machines and equipment in the main stock, %	10.8	32.0	3
10.	Metal content of equipment, tons	1252.0	240.0	5.2

assimilated the production of large fully-equipped block installations for gas fields and their delivery to the site along water ways and over land using air-cushion vehicles.

Taking into account the conditions for developing new gas-transportation and production capacities, booster compression stations with a capacity of 500 thousand kW must be constructed at the Medvezh'e field and booster stations with a capacity of 1.5 million kW must be constructed at the Urengoĭ field. ICPG with production capacity of 20 billion cubic meters of gas per year were built at the Urengoĭ field.

5. Expeditionary–Work-Shift Construction Method

Various forms of expeditionary operations have been used from the very beginning of pipeline construction. As a rule, main pipelines are constructed far from populated points at rapid rates with continuous motion of the construction site, so that the construction work is executed primarily by construction crews that travel to the site. However, previous methods for transporting construction crews from site to site do not address the need to provide the workers and their families with the required social and living conditions. It is for this reason that the concept of expeditionary–work-shift construction was proposed and realized, primarily in assimilating Western Siberian, for outfitting gas fields and laying pipelines.

This method permits interregional utilization of labor resources and social infrastructure: the construction is performed by shifts of mobile construction crews. The workers, engineers, and technicians are settled in camps near the construction sites, and their families live in populated regions, where well-equipped living facilities are available.

Due to the extensive use of interregional labor resources over a period of many years, the organizations in the Ministry of Construction for the Petroleum and Gas Industries has accumulated unique experience in rapid concentration of resources, equipment, and qualified teams for construction of large objects, including also in Western Siberia, where the time available to put the objects into operation is very short.

The construction of the linear part of main pipelines is performed by mobile crews. The method of fully-equipped block construction in gas fields and of compressor stations has permitted extensive use of expeditinary operations.

It is now possible to execute the program of construction and assembly work in Western Siberia based on interregional utilization of labor resources. For example, more than thirty thousand qualified workers, engineers, and technicians from populated regions, who worked for trusts stationed in the European part of the country and in Central Asia, were recruited for the construction of the Vyngapur–Chelyabinsk gas pipeline.

It is projected that in 1985, 110 000 men will be involved in expeditionary–work-shift work in Western Siberia, while by 1990, 150 000 will be so employed, including 80 000 mean from the interregional labor pool.

The socioeconomic and working conditions are continuously improving as a result of regulation of the work routines and remuneration for work performed and priviliges awarded according to the difficulty of the work conditions along the pipeline route, improved living conditions in temporary camps, improved housing and consumer service facilities, and account of the total working time in order to provide workers with the opportunity of resting with their families between shifts. The duration of the shifts, i.e., the total number of days at the construction site, including transit time, is matched with the actual working conditions: seasonal nature of the work, remoteness of the sites, inadequate development of passenger transportation, etc. With partial acclimatization, shift duration must increase with the contrast between the environmental and climatic conditions at the construction site and at locations of permanent residence. The following work and rest cycles were recommended for construction work on northern mainlines: ten weeks on the shift (October–December), three weeks off, followed by ten weeks on the shift (January–March) and 1.5 months off; in less severe regions, 9 weeks on the shift and 1.5 months of independent of season. In laying out gas fields using the work-shift method near locations of permanent residence (for example, laying out fields in the Ob' River Basin) a ten-day shift is possible.

Economic and medical scientific organizations perform systematic research to determine the optimum routines for an extended work day and work week, shift durations for different types of operations, environmental and climatic conditions, and adaptation. Recommendations are formulated from their results.

To perform construction and assembly work in regions with extreme environmental and climatic conditions, special skills are required to operate machines, to weld pipelines and structures, and to perform other technological operations. These shills are acquired over many years of adaptation to the environmental-climatic and living conditions. Laborers, engineers, and technicians, who live in the assimilated regions of the country, but have regularly worked over a period of many years in northern Siberia, can be viewed as specialists, familiar with local conditions.

Their skills and qualifications are entirely adequate for successful completion of work under the specific conditions in the North.

The economic efficiency of the expeditionary–work-shift method is determined primarily by the decrease in the time required to construct main pipelines and outfit gas fields, due to the decrease in the expensive construction of living facilities in the North. Living facilities for a single person in northern regions require four to six times higher capital investment than in the center of the country.

The gas fields in Western Siberia are spread over an extensive territory covered by permafrost and swamps.

Centers of gas production are shifting to the north, into the region that is not favorable for permanent residence and where the period of operation, even though for giant fields, is limited. As a result, the prospects for using labor resources and housing facilities constructed with great effort are uncertain.

Calculations by economists show that extensive application of the expeditionary–work-shift method can save 15–17 million rubles in capital investments and 0.8 million rubles in operating expenses per one thousand workers employed in mainline construction per year. This represents a saving in the form of wages and expenses for maintenance of the camps.

It was determined that to assimilate the oil-gas provinces in the Tyumen region using traditional methods, it would be necessary to settle up to three million people there and to construct upwards of 40 million square meters of housing. Under the coditions of the short times allowed for assimilation, a shortage of labor, and the difficulty of developing an infrastructure in the North, such a program would be unrealistic. For this reason, there was no alternative to the expeditionary–work-shift method of laying main pipelines for assimilating oil and gas fields in Western Siberia. The interregional utilization of labor resources for these purposes can be viewed as an objective reality.

A great deal of attention is directed toward improving the living conditions in the camps used by the work-shift crews near the construction sites and in the settlements along pipeline routes and increasing the comfort of houses and consumer services.

A great deal is also being done to improve living conditions for families, employed by reason of the expeditionary–work-shift method in support cities serving as a base for the expeditionary operations and situated in zones with favorable living conditions, including outside the region.

All of this creates attractive working conditions for the expeditionary–work-shift method in the extreme north. The experimental project-construction office of Minneftegazstroĭ specializes in the development of housing and sanitary-domestic complexes for the camps used by the crews. An entire industry has been created to manufacture housing for the crews, block boilers, water-pumping and sewage treatment plants, as well as dining and sport complexes.

6. Housing Facilities at Construction Sites

Provision of recreational and communal facilities for workers working on gas field construction projects and compressor stations and construction of main pipelines involves prodiguous problems. The construction sites in the gas fields and at compressor stations are situated far from populated points and are cut off from transportation routes. Mechanized columns (production lines) move along the pipeline route very rapidly, which also complicated provision of housing facilities for workers. Under these conditions, temporary structures must be highly mobile and have a high degree of industrialization. In recent years, series production of a complex of such structures has been developed and put into operation.

The VZhK-40 living complex for work-shift crews (Fig. 30) is intended for workers on main pipeline construction projects, as well as for housing the work-shift and repair crews, working on objects for production and transportation of oil and gas.

The complex is assembled from 19 container blocks and consists of the housing and auxiliary parts, connected by the transition block. The housing part consists of a corridor type communal structure with rooms for two people intended to house 40 men. The planar dimensions are 12 x 41 m. The living area is 297 square meters (6.8 square meters per person). The auxiliary part contains a dining hall for 18 people, working on the raw material and the power block. The planar dimensions are 12 x 11.7 m. The building contains a central and three emergency exits. The total area of the building is 573 square meters.

The standardized blocks out of which the building is assembled form a rigid three-dimensional structure consisting of welded panels, which assures that the block are not damaged while being transported over poor roads (the metal used amounts to 99.7 $kg\,m^{-2}$). The external panelling of the block consists of a thin rolled sheet; polystyrene foam is used as insulation; vapor and water barriers are installed; the inner surface of the block is faced with a veneer or plastic over a dry plaster. A heat deflector made of aluminium foil is installed in the floor panels. The windows are tripled glazed. Cold bridges are reduced to a minimum. All of this permits using the blocks and buildings made of them at temperatures down to

Figure 30 VZhK-40 housing complex for expeditionary work shifts.

$-60°C$ in force 3 winds and number 4 snow regions with seismicity up to 9.

The blocks can be transported over railroads and waterways and by motor and air transport. A single block weighs from 6 to 10 tons depending on the equipment with which it is equipped. The dimensions of the block are 12.0 x 2.9 x 2.9 m. The blocks are fully outfitted at the factory with all required sanitation facilities, electricial equipment, and built-in furniture.

The blocks are placed on the foundation by a crane with a weight-lifting capacity of 10 tons with the help of a special cross beam. After the blocks are installed and adjusted, the joints are sealed by prepackaged means and the water supply, heating, ventilation, and power supply systems are interconnected by flexible prepackaged inserts. Grounding, fire alarms, and telephones are installed after the building is assembled. Assembly of the complex requires 147 man-days. The water supply, sewer, power, and heating systems are installed from circuits that are installed inside the foundations. The complex is ventilated by natural and forced circulation. The energy used for heating amounts to 95 000 kcal h^{-1}.

The housing complex for the work-shift crews conforms to degree V fire resistance.

The cost (selling price) of the complex is 196 000 rubles and the erection cost is about 40 000 rubles.

The all-metal standard TsUb-2M block (Fig. 31) is intended to provide housing for workers working on construction of oil and gas pipelinns under conditions existing along the route. The block is distinguished by good mobility, high heat and technical qualities, good sealing qualities, and longevity. Its dimensions are: 9.6 m, diameter 3.2 m, height 3.65 m, and weight 5.8 tons.

The blocks are constructed in the form of a cylinder, formed by a thin rolled sheet wound onto rigid hoops. Polystyrene foam in the form of a shell with a thickness of 100 mm is used as the insulation. The inner facing consists of a wood laminated plastic glued to plywood.

The interior of the block is separated into the following compartments: a lobby which contains a boiler, a lavatory, and a kitchen with an electric stove with two grills; a dining hall; and, a living room. The lobby contains a closet for drying clothes. The block is equipped with special built-in furniture. The windows are triple glazed.

The total area of the block is 27.6 square meters. The living area is 18.9 square meters with a height of 2.6 meters.

The living conditions in the block are made more comforable by installation of a thermal floor: warm air enters from registers situated

Figure 31 All-metal standardized TsUB-2M block.

underneath the floor, resulting in a more uniform temperature distribution over the height of the enclosure. The ventilation system is calculated for thee-fold air exchange. The design of the plumbing and power systems provides for the possibility of connecting the block to external grids.

The blocks are intended for use at temperatures down to $-60°C$, force 3 wind, and number 4 snow regions.

The block are delivered by wheeled carts or sleighs and can be transported by attaching them to a tractor. In addition, they can be transported by all forms of ground, air, and water transport, including by helicopter using an external sling.

The cylindrical shape of the framework gives the following positive qualitites to the TsUB block:

High technological level of manufacturing;

Minimum use of metal ($95\,kg\,m^{-2}$);

Minimum heat radiating surface (the energy consumed for heating amounts to $6000\,kcal\,h^{-1}$);

High strength and transportability;

Streamlining, which is especially important for transport by helicopters to the regions of strong winds and snowfalls.

Figure 32 Route dining hall.

The dining hall for use on pipeline routes provides 22 places (Fig. 32) and is intended for workers on the linear part of main pipelines. The dining hall is intended to service 90 men per hour.

The dining hall is assembled from two blocks with dimensions 9.9 x 3.9 x 3.4 m each. The design is based on the Uyut house cars constructed by the Volokolamsk plant. One of the block contains the dining hall with tables for 22 men, a cafeteria, and an entrance hall with a lobby; the second block contains the store room, refrigerator, kitchen, dishwashing facilities, and a lavatory for the service personnel. The area of the dining and entrance halls is 22 square meters; the area of the kitchen and cafetaria is 30.2 square meters.

The dining blocks are transported on trolleys. They do not require special foundations; each block need only be placed on supports, unloading the springs of the trolleys.

Power is supplied to the dining hall from external sources. In addition, power for heating, lighting, and cooking can be supplied from a diesel generator, which is not included in the complex. The dining hall requires 53 kW. In addition, a gas stove with a gas cylinder is installed in case of power failures in the kitchen.

Figure 33 Sports complex.

Water is supplied from artesian wells or brought in, for which purpose 2000 liter tanks are placed under the ceiling.

The sewage line is connected to a cesspool or to the site main.

Ventilation is by natural convection in all enclosures, but fans are also installed.

The dining hall used on the pipeline route is intended for use under severe environmental and climatic conditions.

The sports complex shown in Fig. 33 is intended for sports and health-improvement activities of workers and employees of construction-assembly crews in remote and inaccessible regions. The sports complex has two halls: a sports hall (Fig. 34) and a hall containing a swimming pool (Fig. 35), as well as auxiliary rooms: a vestibule with a closet for outer clothing, a lavatory, dressing rooms, a first aid room, administrative offices, electric panel and ventilation chamber, and the sports store room.

The sports hall is intended for volley ball, basket ball, gymnastics, badminton, table tennis, wrestling, weight lifting, etc. Thirty people can use the hall at the same time. The hall with the swimming pool is intended for swimming for purposes of health improvement (no competitions and diving). The pool is an Osvod type basin made of a rubberized fabric with

Figure 34 Gymnasium.

pool dimensions 12.5 x 7.5 m and water depth 1.5 m. The pool can handle 15–16 people at the same time. Both halls provide seats for spectators. The sports hall can be used for film presentations and assembly meetings.

The sports complex consists of collapsible fully equipped SKZ type sections. The building has two entrances and its height to the bottom of the foundation is 6 m. The building is erected from collapsible sections with dimensions 3 x 12 m and end panels. The sections consist of enlarged assembly units: double wall panels and a covering panel. The panels are made in the form of a load-carrying steel framework for the building and safety structures. The safety structures consist of triple layer panels with the size of the bearing panels.

Heat is obtained from an external heat network. Water is heated in a separate block boiler. Water temperature after the boiler is maintained automatically with the help of a thermostat. Heating is planned for external air temperatures of − 45°C. The heating system provides the following air temperatures in the enclosures:

Hall with pool + 27°C;

Sports hall + 15°C;

Water temperatures in swimming pool + 24–26°C.

Figure 35 Swimming pool.

The sports complex includes balanced ventilation with forced circulation. During the winter, the system operates with recirculation. The water source can be either a natural basin with pure water or a water supply system from the nearest populated point. Water is discharged into the sewage system for the site. The total installed power for the sports complex is 37 kW. Power is supplied by external power sources.

The medical treatment and health care block Tonus (Fig. 36) is intended for physical therapy and hygienic services for workers in construction–assembly crews working in remote and inaccessible regions of the country.

The block consists of a mobile wagon with dimensions 9.6×3.1 m and height 3 m, placed on a trolley and pulled by motor transport or other means with velocities up to 30 km h^{-1}.

The block includes a room with dry heat (Fig. 37), two showers with water heaters, dressing rooms, a lavatory, a room for relaxation, the pumping unit, and the electrical panel. The dry-heat room contains three-level shelves with removable panels made of aspen boards. In all the other rooms, the walls and barriers are faced with a moisture resistant plastic. The thermal insulation consists of PSB-S insulation and rock wool slabs;

Figure 36 Tonus medical treatment and health care block.

Figure 37 Dry-heat room.

the vapor and water barriers are made of aluminium foil and polyethylene film. The floors in all rooms except the dry-heat room and the showers are made of linoleum on a warm foundation.

Electric heating with a thermostat is installed in the block. The electric stone placed in the dry-heat room provides temperatures up to + 120°C. Natural ventilation is used in all rooms.

The block can be connected to the site power, water supply, and sewage systems. When the block is used under pipeline route conditions, power is supplied from an auxiliary source, while water is supplied from artesian wells or brought in, for which purpose there are two 900 liter storage tanks.

The block can be used with outdoor air temperatures down to − 60°C.

Four to five people can use the Tonus medical treatment block simultaneously.

References

1. Khaĭtun, A. D. (1979). *"Interregional Utilization of Labor Resources", Voprosy Ékonomiki*, No. 8.
2. Bessarab, V. V., A. I. Brun, T. F. Khusnutdinov, V. G. Chirskov, and S. A. Shchetsko (1981). *Stroitel'stvo kompressornykh stantsiĭ v Zapadnoĭ Sibiri [Construction of Compressor Stations in Western Siberia]*, Informneftegazstroĭ, Moscow.
3. Chirskov, V. G., O. Ya. Blech, A. I. Brun, and V. V. Bessarab (1980). *Obobshchenie opyta i analiz stroitel'stva gazoprovoda Urengoĭ–Chelyabinsk [Generalization of Experience and Analysis of Construction of the Urengoĭ–Chelyabinsk Gas Pipeline]*, Informneftegazstroĭ, Moscow.
4. Ivantsov, O. M. (1979). *"Problems of Northern Main Gas Pipelines"*, Izv. Akad. Nauk SSSR, Énerg. Trans., No. 4.
5. Ivantsov, O. M. (1981). *"Component Dependabilities of Gas Transportation", Stroitel'stvo Truboprovodov*, No. 5.
6. Ivantsov, O. M. and A. D. Dvoĭris (1980). *Nizkotemperaturnye Truboprovody [Low-Temperature Pipelines]*, Nedra, Moscow.
7. Ivantsov, O. M., and V. I. Kharitonov (1978). *Nadezhnost' Magistral'nykh Truboprovodov [Reliability of Main Pipelines]*, Nedra, Moscow.
8. Ivantsov, O. M. (1979). *"Scientific Problems Arising in Construction of Northern Gas Pipelines", Stroitel'stvo Truboprovodov*, No. 8.
9. Aĭnbinder, A. B., Yu. P. Baralin, A. V. Val'kovskii, N. P. Vasil'ev, A. M. Zinevich, Yu. V. Kadetov, S. I. Levin, and V. I. Prokof'ev (1979). *Opyt Sooruzheniya Gazoprovoda Vyngapur–Chelyabinsk [Experience Gained in Constructing the Vyngapur–Chelyabinsk Gas Pipeline]*, Informneftegazstroĭ, Moscow.
10. Sedykh, A. D., *O Novom Blochnom Oborudovanii Dlya Nazemnykh Sooruzheniĭ Gazovoĭ Promyshlennosti [New Block Equipment for Surface Construction in the Gas Industry]*, Informneftegazstroi, Moscow.
11. Kashitskiĭ, Yu. A., A. G. Yarmizin, S. A. Lagutin, A. P. Emferenko, A. M.

Lapitskiĭ, A. M. Sirotin, M. S. Fedorov, and Yu. M. Mogil'nitskiĭ (1979). *Oborudyvanie v Blochnokomplektnom Ispolnenii Dlya Avtomatizirovannykh Ustanovok Promyslovoĭ Obrabotki Gaza Po Tipovym Tekhnologicheskim Skhemam [Equipment in Fully-Outfitted Block Form for Automated Industrial Gas Treatment Installations Using Standard Technological Designs]*, Informneftegazstroĭ, Moscow.

12. Rozhkov, A. I. (1980). *Planirovanie Trudovykh Pokazateleĭ S Uchetom Otraslevykh Osobennosteĭ [Projection of Labor Indicators Taking Into Account Special Characteristics]*, Informneftegazstroĭ, Moscow.

Sov. Tech. Rev. A Energy Reviews, Vol. 2, 1985, pp. 289–316
0275–7893/85/002–289 $30.00/0

THE EVOLUTION OF CONTROL PRINCIPLES FOR THE UNIFIED GAS SUPPLY SYSTEM

V. A. SMIRNOV

Institute for High Temperatures, Academy of Sciences of the USSR, Moscow, Korovinskoe Shosse

Abstract

Control principles for the gas supply system of the country have evolved through the enrichment of control concepts, development of a system of plans and forecasts, changing approaches to the problem of reliability of gas supply, and considerable improvement of the principles used to estimate the economic efficiency of the decisions adopted. These and some other paths in the evolution of control principles have been reflected in the evolving industry data control system for the unified gas supply system of the USSR. In this paper, these problems are described in a schematic and consolidated form.

Contents

1. Evolution of Control Concepts

The first point to which we must call attention here is the transition from control of separate elements of the system to control of the system as a whole, constructed on hierarchical principles. This transition was first accomplished for dispatcher control. Later, it entered into the planning and forecasting of the development of the system. It is now being made in the design of system elements: main gas pipelines and gas fields are optimized so as to match local decisions adopted for these elements to the demands of the unified gas supply system (UGSS) as a whole.

The sequence of controlling the gas supply system of the country as a whole entity apparently originated largely from the sequence in which the most important system properties of the gas system had been studied. Thus the most obvious rigid technological coupling of the operational regimes of gas fields, main gas pipelines, and distributing networks, as well as the quite rigid coupling of production and consumption of gas, due to limited possibilities for storing gas, apparently led to the fact that system control of operational regimes was set up first. The study of the wholeness of the gas supply system as an object of planning into the long-term future led to the possibility of forming overall goal functions, restrictions, and so on for the system. In recent years, the concept of the UGSS as some "multidimensional" entity, when the wholeness itself is studied for different aspects, has been developed: economic efficiency, reliability, adaptive capability, etc. [1–4].

The second remark concerning the evolution of control concepts concerns the change in the understanding of the goals of control and the principal methods for forming the system of stylistic developmental and functional goals.

Using the basic assumptions of [1, 3], we can describe the principal control concepts and areas in which they are used as follows.

The Compliance Concept originates from the necessity of obtaining an acceptable variant of the evolution of the system. As is well known, the basic thesis on which this concept is founded is that the systems with which one must deal are extremely complicated. Many aspects of such systems at any stage of evolution will be weakly or completely unstructurable. It is difficult and, in general, at a given stage, impossible even to separate the elements and component parts of the problem according to these aspects and, especially, to describe them analytically in an adequate manner. Under such conditions, the problem is stated so as to obtain an admissible, satisfactory variant or trajectory of system evolution.

This concept corresponds best to statistical approaches to control, extrapolation schemes, and planning of further development from past achievements. The essence of this approach is to provide a statistical description of the trends in past development and to extrapolate these trends with corrections into the future.

The Optimization Concept originates from the need to obtain the best, with respect to some criteria, variant or trajectory of system evolution. The natural prerequisite for applying optimization concepts is the possibility of describing the system and its parts analytically, i.e., here we are talking about well-structured functional and developmental situations when the set of elements and links that must be optimized is completely identified and the entire set has an adequate quantitative description. As is well known, the optimization concept of control corresponds to optimizational methods, which have been extensively developed in recent years both on a general methodological level and in application to gas-supply systems.

The Adaptive Concept is based on the admission that it is extremely difficult to foresee the factors and conditions facing the evolution of the system, as well as a number of problems (information banks) involving the system itself. As a result, the following requirement is established: to provide for a system capability to adapt to the changing conditions of development and internal conditions (in the system itself) with relatively small expenditures on adaptation. Finally, the system must be able to evolve according to a range of variants of future conditions and factors of development and, in addition, an optimality requirement can also be proposed.

The Program-Goal Concept presumes the possibility that the factors and conditions of development conform to the problems and goals of system development. This is what distinguishes this concept from the adaptive concept, where it is assumed that the factors and conditions of development of the system cannot be controlled and the system must adapt to these factors. When full control of the factors and conditions of system development is possible, the problem is stated in terms of the guaranteed achievement of predetermined goals of development under any conditions. Naturally, these goals can be positioned both on the course for selecting the most advantageous variant and on the level of simply adopting the admissible, satisfactory variant.

The unified gas-supply system of the country is a system of unique complexity. However, even here it is possible to identify well structurized aspects (and, in many situations, they have already been identified), for example, material balances in some situations, various kinds of energy-economic and technical-economic characteristics, such as the dependence of costs on the factors that determine them, etc. In planning the development of the UGSS 10 to 15 years into the future, the situation, with respect to many problems, is quite well structured and this has led to the formulation of numerous optimization problems of controlling development by minimizing costs in the presence of a series of restrictions [5]. The same is also true for the Five-Year-Plan method of planning and of several other points of view concerning control.

On the other hand, gas-supply systems are also characterized by control situations that are practically unstructuralizable. These include, for example, long-term forecasts of the development of UGSS 20 to 30 years into the future. Here, optimizational approaches are inapplicable. Satisfactory approaches are possible and quite widely used based on economic-statistical models and extrapolation methods. However, the main path here consists of using complicated, complex concepts of control, about which we will have more to say below.

A weakly structuralizable control situation can also occur with respect to a small part of the system and even in the regime of operational or instantaneous control. If we consider, for example, a compressor station and have in mind not only the production aspect, but also control of the social-psychological climate of the workers and other problems of social control, then this system will become very complicated and practically unstructuralizable for purposes of formulating optimization problems.

In connection with what was said above, if we examine the entire range of the hierarchy of the internal structure of the UGSS, as well as the hierarchy of control problems, then the scale of control problems created is so wide that there is room for all of the control concepts mentioned above. Nevertheless, we can say that control concepts are evolving in the direction of wider application of program-goal approaches and control problems involving adaptation are disappearing. In addition, the program-goal point of view of control has begun to supplement the production and territorial points of view which existed from the very beginning of the creation of the unified system of gas-supply in the USSR.

The most useful areas of application of the program-goal approaches have also been determined. These constitute the spheres of control of scientific-technical programs, regional programs, and programs for improving control.

The development of adaptive approaches involves singling out and studying the basic blocks of uncertainty of information and perturbations which must be considered in controlling the development of the gas-supply system. In terms of the systems analysis approach, the basic sources and blocks of uncertainty and perturbations are subdivided as follows:

Perturbations of the inputs to the system, among which the most important are possible deviations of the actual gas reserves from the planned reserves, as well as possible changes in the supply of resources for development of pipes, aggregates, construction capacities, etc.:

perturbations of the output of the system, consisting of a change in the volumes of gas produced and transported along mainlines; such perturbations are a result of intersectoral corrective adjustments to development plans due to different factors, including changes introduced into the plan for developing gas exports;

perturbations in the system itself, of which the most important ones reduce to deviations of the actual efficiency and time intervals for realizing measures introduced by scientific-technical progress from the planned values.

The latter circumstance is related to the fact that in developing and introducing new techniques and technology, the uncertainty factor looms large and this is an objective property of existing processes.

We shall return to the problem of estimating the perturbations somewhat later. Now we shall examine another problem in the development of control concepts. This is the problem of forming complex control concepts.

One possibility for improving control consists of determining the useful range of control situations for application of each of the control concepts described above. We have briefly examined this path. However, there also exists another path, which could be defended as the more correct path. It consists of forming complex control concepts synthesized from the concepts examined above, which enter in this case as elementary concepts.

This possibility exists since all of the concepts – compliance, optimization, adaptation, program-goal – are interrelated: any one concept can include elements of the others. On the other hand, real control situations always have a complex character: they include aspects that are structuralizable and aspects that are not structuralizable, while causal–uncertainty relations form an integral part of the gas-supply systems themselves, which conform to objective evolutionary laws. The latter indicates that the factors and circumstances of the evolution of these systems can be controlled only partially and, to some extent, they must be included as objectively uncertain factors (at each stage of development).

From this vantage point, the complex control concept is realized in the methods developed in power engineering, and described in [1, 2, 6], for solving problems in the presence of incomplete information. These methods include the optimization and adaptive approaches, the program-goal approach, and the compliance approach (the latter is widely represented by the abundance of heuristic factors in the selection and formulation of the conditions and factors of development, as well as the presence of compliance components in the formulation of the system developmental goals).

Another direction for integration is presented in [3, 4]. It consists of formulating multigoal control of system development based on the identification of special control contours over groups of UGSS system properties, such as economic efficiency, reliability, adaptability, and others. These points of view or control contours supplement the production, territorial, and program-goal aspects of control and are closely interrelated with one another. Control goals, restrictions, and control rules are formulated in each contour. The possibility of formulating a wide range of optimization control problems, including problems of adaptation, reliability, and others, is also considered.

2. Development of the System of Plans and Forecasts

In recent years, the system of plans and forecasts of development of the unified gas-supply system has changed considerably:

The Five-Year Plan has become the main focus of planning (instead of the annual plan), while annual plans are now viewed as stages in the Five-Year Plan;

The requirement that plans be balanced is now extended beyond the Five-Year Plan as a whole and its first year: general and total balance for each year of the upcoming Five-Year Plan and for the Five-Year Plan as a whole are also required;

A ten-year horizon is introduced into planning (in the past, this time horizon was viewed as a forecast);

Long-term horizons for forecasts of development are established: 20 to 25 years into the future; the evolution of the UGSS is now forecasted up to the years 2000 and 2005 (some problems, such as the movement of gas reserves are even extended into a more distant term future);

The role of short-term and current plans and forecasts has also been increased at the same time: they are introduced into continuous planning, and a special organizational system, including the corresponding technical support, etc., is provided for their development.

These evolutionary paths of the sytem of plans and forecasts correspond to the overall directions for improving control. In the meantime, they conform completely to the current unified system of gas supply and its evolutionary trends.

Thus it is well known [6] that the gas-supply system has a very high inertia. Ten to fifteen years are required to make substantial changes in the trajectory along which the system is evolving, since such changes involve corrections to the direction and intensity of development of geological-exploratory operations in different regions of the country, changes in plans for growth of the pipe industry and of other industries supplying needed resources to the national economy. It is well known, in addition, that the inertia of the gas-supply system at the current stage of development is increasing with time [4].

Under such conditions, a critical analysis of ten-year plans in addition to the five-year plans makes planning more interesting, since it permits planning for significant changes in the directions of development and enlarging the freedom of choice in estimating variants of plans. As far as forecasts 10–25 years into the future are concerned, such forecasts have become a necessity, since under the conditions of increasing inertia, development of long-term forecasts provides better justification of the goals for development of the UGSS over a period of ten years, i.e., over the first period of this long-term future. The urgency of long-term forecasting of the growth of the gas-supply system is currently also increasing because power engineering in this country has entered a transitional period in which reliance on fossil fuel as the principal fuel is changing to the use of nuclear power as a primary source [6]. In this transitional period, long-term trends indicating a changeover to efficient utilization of gas in the national economy are appearing, and the structure of scientific-technical progress in production and mainline transportation of gas and so forth will also change. Clarification of the emergent changes is in fact one of the problems of long-term forecasting of the evolution of the gas-supply system.

The significance of short-term plans and forecasts is increasing due to the growth in the scales and complexity of the gas-supply system and as the requirements for synchronous, matched operation of all elements of the UGSS on a common overall load schedule increase.

3. Change in the Approaches Used to Estimate the Reliability of Gas Supply

One of the most important goals of control in the area under study is guaranteeing the required reliability of the gas supply in the national economy. This goal was singled out a long time ago and considerable experience has been accumulated in formulating and solving the corresponding control problems. However, in recent years, it has been recognized that what is usually meant is control of the functional reliability of the gas-supply system via the creation of the corresponding backup and reserves and controls for them. As far as the reliability of the development of the system from the long-term point of view is concerned, control in this area is as yet inadequate, and many control problems are only now being posed [3, 4].

Reliability of development consists of the expected degree of attainment of developmental goals with respect to both the volumes of production and transportation of gas and the efficiency indicators. This category is fundamental to the analysis of the process of adaptation in situations where negative (i.e., hindering control goals) perturbations arise. On the other hand, the process of adaptation with negative perturbations has specific goals and some characteristic properties, and this makes it necessary to single out the special, reliability point of view of adaptation.

To clarify the meaning of the category "reliability of development," we must first emphasize that it does not refer to functional reliability.

The development of the unified gas supply system proceeds by means of construction and reconstruction of objects for production and mainline transportation of gas (including underground storage facilities). For this reason, reliability of development characterizes this aspect of the process: integrated construction and reconstruction of some set of objects in the system. But, in order to realize construction and reconstruction programs, it is also necessary to realize supply programs in other sectors of the national economy. The development process includes these programs as an organic part. The reliability of development thus characterizes the degree of success in introducing into the development of the gas-supply system an entire area of the national economy with extensive cross-linkage between sectors.

The category "negative perturbations," used in analyzing the problems of developmental reliability, is in some sense an analog of the concept "failure," used in estimating functional reliability.

We should keep in mind however that "failure" in the development process refers to the technological processes of construction and reconstruction and the intra- and intersystem linkage which provides for these processes, and it concerns the creation and dissemination of new technology. Developmental "failure" can refer to elements, links in the economic mechanism which ensure fulfillment of the programs for growth of gas supply, etc. This also is what makes the problems and categories of developmental reliability differ considerably from functional reliability.

Finally, we should point out that the means for control of developmental reliability are also specific: this involves formation of a flexible system structure using primarily backup in the supply sectors of the economy rather than in the gas-supply system itself.

We shall briefly examine some of the problems of controlling developmental reliability.

As already noted abovee, one of the large sources of possible perturbations is uncertainty in the information concerning gas reserves. It is shown in [7] that the depedence of the frequency with which deviations from planned reserves appear in actual reserves as a function of the size of these deviations closely conforms to the normal distribution. The density of this distribution increases with the completion of geological-prospecting and geological-exploratory work and, in the future, as fields are introduced into operation as well.

For the adopted (with respect to volume, time intervals, and methods) structure of the geological-prospecting and geological-exploratory work, we can talk about typical (for different categories of reserves) properties of the uncertainty of information. Categories D_1 and D_2 are characterized here by equally probable appearance of different volumes of gas reserves over a quite wide range, while as the transition from the category C_2 to category C_1 and later to categories B and A occurs, the probability density of small deviations of reserves increases. Another aspect of the problem being studied is that large perturbations can occur in the estimates of the sizes of fields to be discovered and their territorial distribution within a given region, just as in perturbations of an interregional nature, involving a change in the entire pattern of distribution of reserves.

In the USSR, the interregional aspect is most important for such large regions as Western Siberia, the European part of the USSR, Central Asia and Kazakhstan, Eastern Siberia, and the marine shelves.

The most reliable increments to reserves in the explored categories are expected in Western Siberia. In Eastern Siberia, large perturbations are expected, since, on the one hand, the predicted reserves there are enormous, while on the other, they are still largely unexplored. In the European part of the USSR, further increments to reserves involve great depths and large perturbations are possible here as well.

In recent years, work has been completed [8] in which the relation between expected perturbations and the amounts of capital invested in geological-exploratory work is analyzed (we are talking about a change from the structures adopted for geological-exploratory work to intensification of such work for goal-oriented control of the uncertainty in the information concerning gas reserves).

Perturbations of the costs and unit consumption of different resources can also be large. Thus estimates show that at the present time the unit technical-economic indicators for different directions of scientific-technical progress in gas transportation can be established with regard for the following ranges of uncertainty (in planning the development of the system up to the end of the century):

30–40% for liquified methane;

15–20% for gas cooled to low temperatures;

10–15% with the use of high pressures (100–120 atm);

5–10% with the use of electrical drive in high-capacity compressor stations.

The more radical a particular direction of scientific-technical progress, the larger the possible perturbations with respect to this direction are. This leads to a kind of stacking of measures for scientific-technical progess as a function of time according to their reliability on the one hand and to the appearance of the category of optimum technology, taking into account the developmental reliability of the gas-supply system on the other [3].

Attempts are also being made to estimate many other types of perturbations that may have to be considered in planning. For this reason, we can say that the amounts of information serving as a basis for formulation of the problem of adaptive planning, more precisely, for planning systems with an adaptive structure, are gradually increasing.

The principal method for controlling developmental reliability is to create flexible structures by means of goal-oriented actions on the corresponding "flexibility factors" [3, 4]. These factors can be subdivided into three groups:

factors for excess and reserves;
factors for structural flexibility;
factors involving the dynamics of costs.

The action of these factors is examined in [4].

We should note that reserves occupy a relatively modest place in the problems of developmental reliability and structural flexibility, while cost dynamics are most important. In addition, as already noted, in order to increase developmental reliability, it is important to create backup capacity in the sectors that provide for the development of the gas-supply system, rather than in the gas-supply system itself. It is this that ensures flexibility when the need arises to change the evolutionary trajectory. Economic reserves in the gas-supply system itself also increase flexibility, but in these manifestations of flexibility, we are talking about management efficiency, which is more important in analyzing problems of functional reliability and less important in analyzing long-term development, since over the long term there "will be time" for reserve capacities of the supply sectors to be used. On the other hand, the cost of creating reserves in the supply sector is much lower than in the gas-supply system.

When developmental reliability is increased, the initial planned costs (i.e., costs that were determined before the appearance of perturbations and not including them) can increase and expenditures on adaptation can decrease. In this connection, it is possible to establish optimum levels of developmental reliability in long-term planning and forecasting (according to calculations, they constitute 0.9–0.98 for some control situations).

The determination of the reliability of development can be justified only for some intensive development of the system, which is determined by an intensive development plan [3]. The category "intensity of development" has approximately the same meaning in reliability control problems as the category of nominal capacity of equipment in functional reliability problems: without establishing developmental intensity levels (just as nominal capacity in estimating functional reliability), it makes no sense to talk about developmental reliability. The categories of developmental reliability and developmental intensity complement one another and can exist only together, and they must therefore be determined simultaneously.

A trajectory of intensive development of the system can be determined starting from the fact that in a situation where computational perturbations appear, the system must use all possible economic maneuvering available to it, i.e., all of the flexibility planned into the sytem. This indicates that optimum reliability and optimum intensity of development coincide [3].

A well-defined set of measures, developed for increasing the levels of reliability of development of the UGSS in the long-term future, is indicated in [4, 5].

These measures include a special policy for distributing geological-exploratory work in order to obtain reliable increments to gas reserves, to create a special "flexible" system structure with respect to selection of a combination of Tyumen and European gas in the gas balance of the European part of the USSR, to create the corresponding backup capacity in the equipment and pipe supply sectors, etc.

4. Development of Principles for Estimating Economic Efficiency of Decisions Adopted

Progress here has largely proceeded along the lines of increasingly more detailed description of the systems properties of the gas supply system and increasingly more complete account of these properties in economic efficiency control problems.

We can arbitrarily single out a number of levels according to the depth of revelation of the systems properties and therefore the quality of control of economic efficiency.

At the first level, this involves only an implicit inclusion of some systems characteristics and properties and control of economic efficiency with a somewhat simplified understanding of this category. In this case, the system is separated out from the background formed by the national economy in such a way that the requirements on the production output of the system. production capacity, and some other output characteristics are established from the point of view of the background economy, thereby ensuring that the requirements for identical energy consumption be satisfied in toto (i.e., equality of the levels of consumption of gas or energy according to all solution variants being compared). The inputs to the system here include capital investments and operational costs, but the requirements imposed by the background economy on the limiting levels of capital investment attracted by the system and the operational expenditures involved are not established (restrictions are not introduced). For this reason, the boundary between the system and the background according to the inputs into the system are not actually indicated.

The basic systems characteristics and properties examined at this level are the cost of the system, production output, and capacity. The economic efficiency is expressed as a rule by some simplified forms of the so-called reduced costs, which include the operational expenses and a special component of the costs, reflecting the relative contribution of capital investments. Some other systems characteristics are also included. Thus the requirements imposed on the parameters of the gas supplied to consumers, systems restrictions on the balance of gas production and consumption (including the regional point of view), restrictions on pressure losses in the system, and other factors are already included at this level of representation of economic efficiency control problems.

The second level of control of economic efficiency is formed when limitations on the permissible levels of some scarce resources, which can form a quite extensive list, attracted by the system, begin to be singled out in detail. This primarily includes restrictions on the use of metal for pipe, gas pumping plant capacities, construction capacities, etc. Naturally, the gas supply system is now more clearly delineated in the background economy.

As experience gained from theoretical investigations [9] and their applications in practice [6] has shown, the concept of economic efficiency is itself greatly altered here. Inclusion of the finiteness of resources has led to the widespread use of "marginal costs"† and other "objectively justified estimates" of resources in problems involving control of economic efficiency [1, 6].

The finiteness of resources together with the heirarchical nature of the unified gas supply system have led to the fact that economic efficiency control problems reflect the problem of matching local and global solutions with the formulation of consistent, multilevel economic efficiency control problems [1, 6, 10].

At the next level, we can examine the approach in which in addition to including the assumptions characteristic of the first two levels, the functional reliability of the system is also accounted for explicitly.

The inclusion of reliability changes the optimal parameters of objects in the gas-supply system as well as the system structure as a whole. Here the problem is formulated as a problem of controlling reserves of capacity within the gas-supply system with a justification of the sizes of these reserves. Excess capacity of objects required for production and transportation of the gas, underground storage facilities, the two-fuel consumer economy, and other factors are viewed as sources of reserves.

Control of economic efficiency in this approach is understood to mean control of the total cost of the system, including the cost of creating a reserve. It is also important to note that, in this approach, the role of several system objects changes considerably. Thus underground storage no longer enters as a means for smoothing irregularities in gas consumption, but as a source for increasing the reserve capacity of the system.

The next, fourth level is characterized by the fact that in singling out the system from the background formed by the national economy, the uncertainty inherent in the inputs and outputs of the system as well as the uncertainty inherent in the internal structural characteristics of the system are also taken into account. This representation of the gas-supply system is based on the fact that the processes involved in its development and functioning are determined not only by causal linkages, but also by a set of complicated uncertain reasons, including random linkages and factors. The boundaries between the system and the background economy in this approach do not define a distinct line, but rather a diffuse region, and this concerns both system inputs and outputs.

The diffuseness of the boundaries and internal characteristics of the gas-supply system form a zone of uncertainty in the states of the system

†Translator's note: The use of the term "marginal" in this context is explained in the translator's note in the paper by

[1, 6]. which ultimately change in a fundamental way the essential idea of economic efficiency control: a category called "zone of equally economic solutions" [1, 3, 6] appears and the rules for "probing" the zone of uncertainty and adoption of solutions in it are formulated. In addition, the functional reliability of the system as a system category now expresses part of the problem of establishing interrelationships between the planned system and the national-economic background under conditions of uncertainty (incomplete information).

The fifth level, supplementing the preceding level, is characterized by the following basic features:

Attempts are made to include explicitly the interconnectedness of the states of the system and the national-economic background along the development trajectory: to single out and include the characteristics of system inertia and flexibility in the development process and to include reaction and feedback factors of future states of the system and the background on their instantaneous states;

Reliability requirements imposed by the background on the system and by the system on the background are singled out not only with respect to system functioning, but also with respect to the system development, and the systems requirements for developmental reliability are introduced;

A cause–effect relation is established between the levels at which uncertainty appears at the input and output of the system. Depending on the flexibility planned into system, the system can extinguish to some extent the perturbations at the input and it can make the outputs more stable.

In short, an attempt is made here to take into account as completely as possible the different system states in controlling the economic efficiency of the system during its development [1, 3, 6].

In so doing, the concept of economic efficiency itself is further developed. For example, it is noted in [3], along these lines, that the collection of control variables changes, optimization takes into account a range of conditions, and contradictions between the local optima of different states of the same system are formed in an explicit form: the closer the gas supply system is to an optimum for one of the extreme states of the resource limitation vector, the farther it will be from the optimum in the other extreme state of the resource vector. A special problem of matching local optimum states of the system under different conditions fixed by the background arises.

The presentation above does not exhaust all possible evolutionary paths for methods of controlling the unified gas supply system, but it does include the most important ones.

In what follows, we shall examine the manner in which the development of control methods was reflected in the structure and essence of the evolving automated system for controlling the gas industry.

The first automated control systems (ACS) were created in the gas industry in the middle of the 1970s. The introduction of ACS was preceded by a large amount of work on restructuring the organizational scheme for controlling the gas industry and creating an industry network of computing centers. The management structure of control, existing in the middle of the 1970s, was distinguished by its cumbersomeness and many links, and it did not permit efficient use of modern computer technology. The new scheme for controlling the industry provided for the transition to a two- (ministry–industry association) and three-link (ministry–all-union industry association–production association) structure and included the following characteristics of the present stage of development of the gas industry:

Displacement of production centers from the European part of the country into regions with difficult conditions for organizing labor and control;

High rates of growth of gas production in the country;

Multifaceted nature of the work of the Ministry of the Gas Industry (well drilling, production, transportation, refining, and distribution of gas, production of reserve parts and repair of technical equipment, production of gas equipment, etc.);

Connection of production, refining, transportation, and utilization of gas.

The organizational restructuring of industry control created favorable conditions for introducing ACS at all hierarchical levels.

However, in the middle of the 1970s, the system properties of the gas industry as an object of control were not adequately studied.

For this reason, within the scope of ACS, the compliance aspects of control were primarily realized first. Automation was of a local nature, since computers were used to solve separate problems in planning and the inclusion and analysis of production-economic activity.

The first ACS, created at different levels of the control hierarchy, were weakly interconnected, which decreased their efficiency.

Meanwhile, experience in operating the first ACS yielded valuable information for the study of the systems properties of the gas industry. This permitted introducing the second generation ACS, based on the optimization methods of control, into the gas industry already in 1980 [11]. Second generation automated control systems involve the use of interrelated criteria of optimization and numerous internal (between elements of the system) and external (with adjacent sectors and with the energy complex) links of the unified gas supply system were taken into account. Since it is impossible to determine uniquely the conditions under which a plan will be realized, the development of the UGSS was optimized taking into account the requirements of the adaptation concept. The introduction of optimization methods of control was accompanied by system coordination of ACS functioning at different levels of control.

This created favorable conditions for creating a unified. hierarchically correlated, complex of automated control in all links of the organizational structure of the gas industry.

The reasons for introducing this complex are as follows:

To improve the economic mechanism and to increase the efficiency of industry planning based on the use of economic-mathematical methods and computers [12, 13];

To increase management efficiency, maneuverability, and reliability of control of the unified system of gas supply [14]; and,

To increase the efficiency of technological processes involved in the production, refining, and transportation of gas based on complex automation and the use of computers [15].

The structure of the automated control system for the gas industry (ACSG) includes:

A sectoral automated control system of the gas industry (SACSG) together with the automated dispatcher control system of UGSS entering into it (ASDC UGSS);

ACS of all-union industry associations (ACS AIA);

ACS of production associations (ACS PA).

The functioning of the automated control systems at different hierarchical levels is provided for by a developed sectoral network of information-computer centers within the gas industry together with the means of collecting and transmitting information.

The industry network of information-computer centers represents a unified complex, including the main computing center (MCC) of Mingazprom, clustered information-computer centers (CICC) of AIA, and information-computer centers (ICC) belonging to the production associations.

The structure of the complex indicated corresponds to the structure of ACSG, forming a sequential chain of information links MCC–CICC–ICC.

Information stations, intended for collecting and transmitting technological and economic information to the appropriate level of control, have been created in plants for complex preparation of gas, head structures in gas fields, main gas pipelines, and underground gas storage facilities.

The automated control systems entering into ACSG form two functional paths:

Regime-technological (ACS of the unified gas-supply system);

Organizational-economic (automated system of planning and accounting in the gas industry (ASPAG).

The regime-technological path ensures automation of control of UGSS and includes ASDC UGSS at the top level, and ACS for technological processes at the level of unification and separate objects. The organizational-economic path deals with the problems of automating long-term, five-year, and annual planning, accounting, operational control, and analysis of all levels of control in the gas industry.

The upper level of ASPAG consists of the organization-economic subsystem of SACSG, the middle level includes the economic-organizational subsystems of ACSAIA, and the lower level includes the economic-organizational subsystems of ACS PA.

We shall examine in more detail the automated system of planning and accounting in the gas industry (ASPAG). The purpose of this system is to increase the efficiency of control by improving planning based on the use of economic-mathematical models and more efficient computer analysis of the large volumes of planning-accounting information.

ASPAG must evolve in two stages.

At the first stage (1981–1985), organization-economic ACS at different levels of control will be functionally and informationally linked, unification of data bases will be completed, and a unified system of linear optimization models for annual, five-year, and long-term planning will be introduced. At the end of the first stage, the transition to the program-goal control concept will begin.

At the second stage (1986–1990), the creation of a multilevel integrated ASPAG, based on a data bank distributed over the levels of the hierarchy, transfer between computers, and a unified system of nonlinear dynamic models ensuring continuity of planning will be completed.

For the conditions under which ASPAG functions, automatization must provide for the following:

improvement of the technology of planning and the process of forming long-term, five-year, and annual plans at all stages of their development based on the use of economic-mathematical methods and computers;

more complete planning due to closer coupling of sectoral, territorial, and program viewpoints of the plan;

balance of the plans for production, transportation, and distribution of gas in order to increase the reliability of gas supply, as well as coordination of the main production, capital construction, and material-technological provisions;

better substantiation of the adopted plans through the use of optimization multivariant calculations at all levels of control.

The planning and accounting functions over the basic forms of activity of the sector (production and transportation of gas, capital construction, etc.) are automated with the help of the complex of models.

Within the framework of ASPAG, the optimum variants for development of fields, ensuring increased production with minimum expenditures of material-technical and financial resources, are determined and gas production is coordinated with the volume of well drilling and the increase in explored reserves and throughput capacity of gas pipelines.

In forming the projected plan, economic-mathematical models are used at all levels of control. These models permit establishing the maximum permissible volume of gas production, including geological-technical characteristics of fields and their actual state, the dynamics of change in gas reserves, the capacity of drilling organizations, and the state of provision of the fields.

When automating within the framework of ASPAG, certain characteristics of gas production planning must be taken into account.

Long-term planning of gas production is conducted starting from existing raw material resources and geological-exploratory data on the increment to gas reserves in the next ten years.

The five-year production plan includes the real possibilities of the fields being developed and data on the possible increment to gas production from new fields.

A sector plan for gas production, which realizes the control numbers of the state plan of the USSR for the planning period with minimum monetary expenditures, is formulated at the ministry level with the help of economic-mathematical models of SACSG. In so doing, the production capabilities of AIA, restrictions on material-technical resources that have been singled out, the capacity of gas transportation systems, and the capabilities of the construction organizations are taken into account.

Modeling yields the optimum industry plan for introducing new production capacity in the fields and a plan for conducting the drilling operations for gas, oil, and condensate. Computer calculations of capital investments in provisions for gas and gas-condensate fields, as well as in exploratory and operational drilling, are performed at the same time.

The problem of the distribution of gas production over the gas fields is solved within the framework of ACS AIA in such a way that the planned volume of gas production for AIA is guaranteed with minimum monetary expenditures including restrictions following from the technology for developing the field and the time periods required for put objects of capital construction into operation.

Under the conditions of large distances between the source of raw material and the principal gas consumers, improvement of planning and accounting plays an important role in gas transportation.

A system of economic-mathematical models has been developed for automating the planning of development of UGSS for ten-, five-, and one-year intervals [16–19]. Models of the five-year plans of gas transportation with tasks distributed by year, which is due, as noted above, to the leading role of five-year plans in the entire system of economic planning, play the central role in the system indicated. For a long planning period, the five-year period is a guide, relative to which variants of the development of the gas supply system into the long-term future, as well as the annual planned requirements, are evaluated. This approach ensures continuity and consistency of planning at different stages in time.

In long-term planning, basic trends and prospects for development of mainline gas transportation in ten-year periods are determined within the framework of ASPAG (with the fifth and tenth years singled out as computational levels). At this stage, new directions of gas delivery to consumers from production regions are justified, variants of efficient loading of operational gas pipelines are determined with regard of the possibility of introducing new intersystem ties, and the efficiency of new solutions to technological problems for realizing the network of interregional gas flows is evaluated.

The main solutions, related with the choice of new paths for introducing gas and efficient loading of operating gas pipelines, are reflected in the network of gas flows along the unified gas-supply system over a ten-year period with the indicators for the fifth and tenth years singled out.

For the conditions under which ASPAG functions, as efficient network of gas flows with long-term planning is constructured with the help of economic-mathematical models.

In the simulation, the UGSS is represented as a network consisting of nodes and arcs. The nodes represent fields, consumers, underground gas storage facilities, and key compressor stations. The arcs represent sections of operating and possible new gas pipelines.

Each node is characterized by the volume of production or consumption of gas. To determine the gas consumption volumes, a complex of problems, which take into account and analyze on a computer the solutions concerning future utilization of gas, as issued by the state plan of the USSR, is used to determine the gas consumption volumes.

In the USSR, gas belongs to the class of products that are subject to centralized distribution.

Each year, based on plans of future production volumes and norms of gas consumption for delivery of unit production, the state plan of the USSR issues permission for utilization of natural gas by enterprises in the different ministries and departments that are in the process of being constructed, enlarged, and converted from the use of other types of fuel.

The complex of problems involved in taking inventory of and analyzing authorizations for the use of natural gas is dealt with by automating the collection and processing of information furnished by the state plan of the USSR, as well as by determining the increment to the consumption of gas for the planned period broken down according to systems supplying gas to cities, provinces, republics, economic regions, sectors in the national economy, ministries, and departments taking into account the trends in gas consumption, fuel performance, and types of interchangeable fuel, fuel regimes and forms of substitution fuel.

Starting from the planned volumes of production and consumption for a ten-year period, an optimal scheme for interregional gas flows is formulated.

In modeling each arc, the possible directions for transporting gas, section length, restrictions on the throughput capacity of operating gas pipelines, and the technical-economic characteristics of gas transportation are compared. For arcs reflecting possible directions of gas flow, restrictions on throughput capacity are not given. The technical-economic characteristics of gas transportation are used in constructing the functional of the model.

The technical-economic characteristics of operating and possible new gas pipelines are distinguished [20].

The technical-economic characteristics of operational gas pipelines constitute the dependence of operational costs on the volume of gas transported. Operational costs for each gas pipeline must be calculated starting from the specific realization of the linear part, the spacing between the compressor stations, and the scheme for connecting gas pumping plants at compressor stations.

The technical-economic characteristics of gas pipelines being constructed express the dependence of the reduced expenditures on the volume of transported gas.

They are constructed for the basic standard sizes of pipes used in the industry (diameter 720–1420 mm), for supercharger pressures of 5.6 and 7.5 MPa, and a wide range of variation in the strength of pipe metal.

Each standard pipe-size is matched to a set of standard gas pumping aggregates and schemes for connecting them. The simulation of characteristics for a possible realization of the linear part of the gas pipeline and of a compressor station is determined (with varying volume of transported gas) by the length of the computed section and reduced expenditures (related with construction of one km of the given pipeline).

Using the technical-economic characteristics of gas transportation, a scheme of interregional gas flows along UGSS over a ten-year period is constructed within the framework of ASPAG. The economic-mathematical model permits finding for the fifth and tenth years the volume of gas transported along sections of operating gas pipelines, as well as new directions and volumes of gas flow along them, ensuring that minimum reduced costs are achieved during the entire period of the plan [19].

In so doing, the inflow, production, utilization, and transmission of gas must be balanced at each node of the UGSS. The boundary conditions are the throughput capacity of the sections of operating gas pipelines, production and consumption volumes, and volumes of gas that are imported and exported.

For long-term planning, the starting information may not be uniquely specified. Usually, sets of starting data are prepared, differing by production volumes and their distribution over different gas fields, consumption volumes (within the limits permitted by the state plan of the USSR), and variants of development of the gas supply network. The so-called optimal standard solution is obtained from calculations performed with each set of initial data. This solution is strictly optimal only for given conditions and the adopted criterion for optimization. In addition, planning of interregional gas flows has a multicritierion character: together with the criterion of reduced expenditures, a number of additional factors must be included, such as the expedient organization of construction operations, reliability of gas-supply schemes, informal limitations, etc.

For this reason, the set of optimal standard solutions obtained is analyzed by specialists, who take into account additional restrictions and propose new schemes for feeding gas [11].

The accumulated experience in man–machine dialogue, for comparative analysis of different alternatives of development of UGSS, permits obtaining within an acceptable period of time a so-called compromise solution, which is recomended as the scheme for gas flows [16].

The solutions obtained at the long-term planning stage from the foundation for formulating a plan within the framework of ASPAG for the upcoming five-year period with tasks broken down by year.

A characteristic feature of the five-year planning process for gas transportation is the presence of limitations on the material-technical resources of the sectors singled out.

In the presence of limits on some types of resources, the criterion of optimality in the problem of determining interregional gas flows depends not only on the volume of gas transported, but also on the means chosen for technical provision of possible new gas pipelines.

In this connection, the mathematical formulation of the problem in general form reduces to a complicated model of nonlinear discrete-continuous programming [19]. Methods have been proposed within the framework of ASPAG for decomposing the indicated model into continuous and discete parts, corresponding to the problems of determining the gas flows and their technological realization.

The mathematical model permits determining for each year in the Five-Year Plan a flow distribution along UGSS (including possible new directions of gas flow) which ensures minimum reduced expenditures on gas transportation within the five-year plan as a whole under the condition that balance relations are satisfied at the nodes of the network in each year. In so doing, the throughput capacities of the operating gas pipelines at the beginning of the period and the restrictions on the development of the system within the period of the plan must be taken into account, while the state of the UGSS at the end of the five-year plan is matched with the decisions adopted for ten years into the future.

For most sections of developing gas pipelines, a requirement of non-decreasing (according to the years of the five-year plan) gas flows is introduced into the model because otherwise the costs of switching and reorganizing construction subdivisions increase sharply [16, 19].

As a result of the modeling, the volumes of gas transported and the directions of gas flow from the gas fields are determined for sections of the UGSS by year within the five-year period. This information serves as a basis for calculations, performed within the framework of ACS UGSS, of variants of the technical realization of interregional gas flows, providing for delivery of the assigned quantities of gas to consumers while maintaining the required pressure at the nodes of the network and minimizing the costs of expanding operating gas pipelines or constructing new ones.

The parameters of each pipeline and selected with regard of the staged nature of its development. In this approach, the pipeline parameters are first calculated assuming complete development with the help of mathematical models. This solution is taken as an upper limit when optimizing the technical realization by year within the planning period. In so doing, the costs over the entire period of development of the gas pipeline are minimized [21].

Based on the calculations indicated, different variants of the introduction of new capacity on main gas pipelines are determined by year for the planning period. Within the framework of the subsystems of ASPAG, which automates the control of capital construction, the new capacities indicated are compared with the limits on pipe, gas pumping plants, and other equipmet and the scheme of gas flows along UGSS is then refined.

Starting from the introduction of new capacity, ensuring the best realization of the adopted scheme of interregional gas flows under conditions of finite material-technical resources, the sequence of construction of new objects is determined.

The mathematical model permits establishing for each object the times for starting construction that guarantee minimum total reduced costs over the entire planning period, while conforming to the basic restrictions on the material-technical resources and capital investments and taking into account the assignments for introducing new production capacities as well as the possibilities of construction organizations. These calculations serve as a basis for the formulation of plans for capital construction in the planning period.

Based on the flow schemes, an industry plan for gas transportation by year within the five-year period is calculated on a computer and the volume of freight transportation and the basic economic indicators are determined.

With the help of the complex of problems, the industry automated control system is used to formulate the five-year plan for transportation of gas put out by the Ministry of the Gas Industry (Mingazprom) of the USSR. This plan contains summary data for each year of the planning period as well as consolidated indicators for AIA and PA (without breakdown by sections of the system of gas pipelines).

This information is presented in the form of control numbers, which must be elaborated in five-year plans for gas transportation, developed on a computer within the framework of ASPAG at the AIA and PA levels.

Automation of the production-economic work of AIA over the five-year period provides a gas transportation plan that guarantees the realization of the control numbers generated by Mingazprom with minimum monetary expenditures. Using the indicators of the gas transportation plan put out by Mingazprom as boundary conditions, the internal needs and losses of gas and the volumes of gas consumed in provinces and cities, broken down by sections of gas-pipeline systems, are refined within the ACS AIA and the scheme of gas flows along the gas supply network in AIA for the period of the plan is also calculated.

Special attention is devoted here to the technological realizeablity of the indicated flow schemes, making sure that the required pressures are maintained at the nodes of the gas-supply network.

For this purpose, an interaction is set up between the organizational-economic and regime-technological subsystems of ACS AIA.

The distribution of gas consumption along the sections of the gas pipeline system is adjusted (within the limits permitted by the control numbers from Mingazprom) according to the results of the regime-technological calculations and the gas flow scheme is refined.

Based on the scheme of gas flows along the AIA gas supply network, the AIA transportation plan, which includes the indicators for subordinated associations broken down by sections of the gas pipeline system in addition to summary data, is formulated in a form showing the breakdown by year within the five-year period with the help of economic-mathematical models. The calculation of the five-year gas transportation plan is supplemented by a determination of the basic economic indicators corresponding to it for the period of the plan (volume of freight-transport work, cost, profit, volume of realization) for AIA and PA.

Starting from plans formed at the AIA level, within the framework of ACS PA, the indicators are broken down according to the linear production controls of main gas pipe lines and the volumes of gas supplied to consumers are refined by year within the planning period.

The plan indicators, determined at the five-year planning stage, are worked out in detail in the annual planning.

An annual gas-transportation plan (with breakdown by quarters), which makes possible the completion of the tasks set forth in the plan with minimum monetary expenditures and ensures that the required balance is achieved, is formulated within the framework of the industry ACS [17, 22]:

total volumes of gas produced and consumed by Mingazprom;

basic gas transportation indicators within the framework of PA and AIA;

overflows between neighboring PA taking into account the capacity of gas pipeline systems;

planning indicators of Mingazprom, AIA, and PA with balances and gas-distribution plans from the state plan of the USSR.

Concolidated indicators, contained for each association in the annual gas-transportation plan constructured by Mingazprom, enter into ACS PA for use as control figures.

Relying on the indicated figures and the gas stocks set aside for consumers in the association, the indicators of the annual transportation plan for gas produced by the production association are determined within the scope of ACS PA.

The problem of automating the planning of gas transportation with the help of the ACS AIA economic-mathematical models is formulated as follows: starting from the control figures provided by Mingazprom and the analysis of PA, the problem is to formulate a gas transportation plan for AIA, taking into account the functional characteristics of the gas supply network of the region, providing at lowest cost a reliable flow of gas to consumers in volumes provided for by the allotted gas stocks.

At all levels of control (Mingazprom, AIA, PA), the formulation of the annual gas transportation plan within the framework of ASPAG is supplemented by a calculation of the cost, volumes of realization, profit, and freight-transportation volume.

In the process of fulfilling the annual and quarterly gas transportation plans, information concerning the progress of the production-economic activity must be exchanged between the associations and the Ministry.

An account of the fulfillment of the gas transportation plan is delivered to Mingazprom monthly by AIA and PA. Automation of acounting and analysis of gas transportation permits formulating accounting information within the framework of ACS at lower levels and transmitting this information to organizations at higher levels for purposes of economic analysis with the help of information-logical problems of ACS AIA and SACSG. Based on data on the fulfillment of the gas transportation plan, entering along communication channels at the Mingazprom level from AIA and PA, summary accounts, different kinds of reports, and an analysis to elucidate the reasons for any deviations from the planned tasks and timely execution of appropriate controlling actions are formulated.

The complex of models for monitoring, accounting, analysis, and operational control provides the workers at the Ministry with timely and reliable information, required for controlling the realization of construction plans and the plans for putting new production capacity into operation.

In addition, the complex indicated automates the process of obtaining a system of factual indicators characterizing the following:

Rates of growth and increments to the production capacities and basic stocks, capital investments and construction-assembly work, changes in the structure of capital investments in the basic operations of the industry, time intervals required for construction and assimilation of new production capacity, volumes of transitional and newly initiated construction, and the volume of construction in progress, and other factors;

Reasons for increase or decrease in the volumes of gas produced, transported, and stored, metal consumed, unfinished construction, and others;

Effectiveness of basic stocks and capital investments both for the industry as a whole and in separate subsectors.

Current planning and accounting are supplemented by monthly, weekly, and daily operational control of the operation of the UGSS. These problems fall outside the scope of this review and are examined in detail in [23–26].

The use of economic-mathematical methods and computers for purposes of planning and control makes it easier for the industry to achieve high quality final results.

Because it optimizes the current and long-term plans, balanced with the allotted resources, ASPAG is an important tool for improving the operation of the economic mechanism in the gas industry.

References

1. Melent'ev, L. A. (1979). *Sistemnye Issledovaniya v Énergetike (Systems Studies in Power Engineering)*, Nauka, Moscow.
2. Melent'ev, L. A. (1977). *"Principal Resources of Large Power Systems"*, *Izv. Akad. Nauk SSSR, Energetika i Transport,* No. 1.
3. Smirnov, V. A., S. V. Gerchikov, and V. G. Sokolov (1978). *Otsenka Nadezhnosti i Manevrennykh Kachestv Planov (Evaluation of the Reliability and Flexibility of Plans)*, Nauka, Novosibirsk.
4. Smirnov, V. A. (1981) *"The Concept of Long-Term Development of the Unified Gas Supply System of the USSR", Izv. Akad. Nauk SSSR, Energetika i Transport,* No. 4.
5. Smirnov, V. A. (1979). *"Pipe Transportation Systems", Izv. Akad. Nauk SSSR, Energetika i Transport,* No. 4.
6. Makarov, A. A., and A. G. Vigdorchik (1979). *Toplivno-Énergeticheskiĭ Kompleks (The Fuel-Energy Complex)*, Nauka, Moscow.
7. Volkonskiĭ, V. A., T. A. Kosenko, V. A. Smirnov, and I. Ya. Faĭnshteĭn (1973). *"Probabilistic Evaluation of the Increment to Gas Researves", Gazovaya Promyshlennost',* No. 4.
8. Kitaĭgorodskiĭ, V. I., and I. Ya. Faĭnshteĭn (1981). *"Problems of Reliability of the Raw-Material Provisions for the Production of Natural Gas", Izv. Akad. Nauk SSSR, Energetika i Transport,* No. 4.
9. Aganbegyan, A. G. (1979). *Upravlenie Sotsialisticheskimi Predpriyatiyami (Control of Socialist Enterprises)*. Ékonomica, Moscow.

10. Smirnov, V. A., and S. V. Gerchikov (1978). *"Planning Flexibility in Two-Level Optimization"*, in: *Modelirovanie Vnutrennikh i Vneshnikh Svyazeĭ Otraslevykh Sistem (Simulation of the Internal and External Links of Sectoral Systems)*, Nauka, Novosibirsk.
11. Borozdenkov, A. I., A. S. Firer, and B. L. Tsypin (1980). *"Problem Complex for Level II Management-Economic Subsystems of SASC", Obzorn. Inform. "Avtomatizatsiya, Telemekhanizatsiya i Svyaz' v Gazovoĭ Promyshlennosti (Review: Automation, Telemechanization, and Communication in the Gas Industry)*, VNIIÉgazprom, Moscow.
12. Margulov, R. D., and E. K. Selikhova (1978). *"Methodology of Planning for the Unified Gas Supply System: Status and Problems", Gazovaya Promyshlennost'*, No. 11.
13. Selikhova, E. K., and É. B. Bagieva (1980). *"On the Way to Improvement of the Economic Mechanism of the Gas Industry", Gazovaya Promyshlennost'*, No. 4.
14. Khalatin, V. I., and M. M. Maĭorov (1982). *"Improvement of the Control of the Unified Gas Supply System of the USSR Based on the Introduction of Computer Technology", Gasovaya Promyshlennost'*, No. 5.
15. Maĭorov, M. M. (1980). *"Paths for Further Increase in the Efficiency of Automation and Utilization of Computer Technology in the Gas Industry"*, in: *Obzor. Inform. Avtomatizatsiya, Telemekhanizatsiya i Svyaz' v Gazovoĭ Promyshlennosti (Reviews: Automation, Telemechanization, and Communication in the Gas Industry)*, VNIIÉgazprom, Moscow, No. 1.
16. Zhuchenko, I. A. (1978). *"Models of Optimization of Planning of Gas Flows", Gazovaya Promyshlennost'*, No. 11.
17. Firer, A. S., and V. I. Sheremet (1978). *"Methematical Simulation of the Gas Transportation Plan", Gazovaya Promyshlennost'*, No. 11.
18. Stavrovskiĭ, E. R., and V. A. Efremov (1979). *"Inclusion of Consumption Dynamics in Optimizing Plans for Development of the Unified Gas Supply System", Gazovaya Promishlennost'*, No. 12.
19. Zhuchenko, I. A., V. I. Feĭgin, E. P. Frolova, T. I. Shtil'kind, and E. E. Mikhaĭlova (1981). *"Optimization of Gas Flows in the Unified Gas Supply System for Future and Mid-Term Planning"*, in: *Obzor. Inform. Ékonomika Gazovoĭ Promyshlennosti (Reviews: The Economics of the Gas Industry)*, VNIIÉgazprom, Moscow, No. 4.
20. Vaserman, V. O., and A. S. Firer (1981). *"Construction of Technical-Economic Characteristics of Gas Transportation and Their Use for Optimizing Interregional Gas Flows"*, in: *Obzor. Inform. Ékonomika Gazovoĭ Promyshlennosti (Economics of the Gas Industry)*, VNIIÉgazprom, Moscow, No. 4.
21. Galustova, L. A., N. Z. Shor, N. E. Rozen, A. Sh. Kardonskaya, and A. I. Momot (1978). *"Optimization Algorithms for Solving the Problems of the Development of the Unified Gas Supply System", Gazovaya Promyshlennost'*, No. 11.
22. Antonov, M. K., A. S. Firer, and V. I. Sheremet (1981). *"Improvement of Current Planning of Gas Transportation Based on the Use of Computers"*, in: *Obzor. Inform. Ékonomika Gazovoĭ Promyshlennosti (Economics of the Gas Industry)*, VNIIÉgazprom, Moscow, No. 4.
23. Berman, R. Ya., and V. S. Pankratov (1978). *Avtomatizatsiya Sistem Upravleniya Magistral'nymi Gazoprovodami (Automation of Control Systems for Main Gas Pipelines)*, Nadra, Lenigrad.
24. Khalatin, V. I. (1979) *"Dispatcher Control in the Unified Gas Supply System"*, in: *Obzor. Inform. Transport i Khranenie Gaza (Review: Transportation and Storage of Gas)*, VNIIÉgazprom, Moscow, No. 7.
25. Sukharev, M. G., E. R. Stavrovskiĭ, and V. E. Bryanskikh (1981). *Optimal'noe Razvitie Sistem Gazosnabzheniya (Optimal Gas-Supply Systems)*, Nedra, Moscow.

26. Kuchin, B. L. (1979). *Operativnaya Informatsiya v ASU Magistral'nykh Gazoprovodov (Operating Information for the Automatic Control System of Main Gas Pipelines)*, Nadra, Moscow.